Engineering Principles
Volume 1

Electrical and Electronics Technicians Series

Consulting Editor C. Child, C. ENG., MIEE, MIERE

Wandsworth Technical College
Member, Syllabus Sub-committee and Examiner for City and Guilds Course 57

Mathematics Volumes 1 and 2
J. W. Davies

Engineering Principles Volumes 1 and 2
T. A. Lovelace

Engineering Principles

Volume 1

T.A. Lovelace
Carshalton College of Further Education

Nelson

Thomas Nelson and Sons Ltd
Lincoln Way Windmill Road Sunbury-on-Thames
Middlesex TW16 7HP
P.O. Box 73146 Nairobi Kenya

Thomas Nelson (Australia) Ltd
19-39 Jeffcott Street West Melbourne Victoria 3003

Thomas Nelson and Sons (Canada) Ltd
81 Curlew Drive Don Mills Ontario

Thomas Nelson (Nigeria) Ltd
8 Ilupeju Bypass PMB 1303 Ikeja Lagos

First published in Great Britain 1970
Reprinted 1974; 1975 (with updated symbols), 1976

Copyright © T. A. Lovelace 1970

Illustrations by Colin Rattray

0 17 741108 2

Printed in Hong Kong

Contents

Preface ix

List of Symbols and SI Units xii

1 The International System of Units
1.1 Basic units **1**; 1.2 Derived units **2**; 1.3 Distance–time graph **3**; 1.4 Velocity–time graph **4**; 1.5 Mass **6**; Additional questions **8**

2 Newton's Laws of Motion
2.1 The law of inertia **9**; 2.2 The law of momentum **9**; 2.3 The reaction force law **12**; 2.4 Friction force **13**; Additional questions **14**

3 The Effects of Force
3.1 Stress **15**; 3.2 Types of stress **16**; 3.3 Strength of materials **17**; 3.4 Strain **18**; 3.5 Relation between stress and strain **19**; 3.6 Factor of safety **21**; Additional questions **21**

4 The Pictorial Representation of Forces
4.1 Scalar quantities **23**; 4.2 Vector quantities **23**; 4.3 The use of vector diagrams **24**; 4.4 The theorem of the triangle of forces **26**; 4.5 Centre of mass **30**; 4.6 Practical applications of the theorem of the triangle of forces **30**; 4.7 Polygon of forces **34**; 4.8 Resolution of forces into right angle components **37**; 4.9 The effect of the weight of a mass resting on an inclined plane **38**; Additional questions **40**

5 Moments and Parallel Forces
5.1 Definition of a moment **45**; 5.2 Bodies in equilibrium **46**; 5.3 Reaction at the pivot **46**; 5.4 Beam supports **50**; 5.5 Centre of mass **50**; 5.6 Calculation of the reactions at the beam supports **51**; 5.7 Parallel forces **53**; 5.8 Torque **55**; Additional questions **56**

6 Work, Energy, and Power
6.1 The meaning of the term 'work' **58**; 6.2 Energy **58**; 6.3 Power **63**; 6.4 Efficiency **63**; 6.5 Practical units of power **64**; 6.6 Practical units of work **64**; 6.7 The cost of energy **65**; 6.8 Work done by a variable force **65**; Additional questions **67**

7 Machines
7.1 The machine **69**; 7.2 The lever **69**; 7.3 Velocity ratio **70**; 7.4 Mechanical advantage **70**; 7.5 Efficiency of a machine **70**; 7.6 Lever

systems **71**; 7.7 The inclined plane **73**; 7.8 Pulley blocks and tackles **74**; 7.9 The differential axle **77**; 7.10 The wheel and differential axle **78**; 7.11 Gearing **79**; 7.12 The winch **81**; 7.13 The worm and wheel **82**; Additional questions **84**

8 Rotary Movement

8.1 The measurement of angles **86**; 8.2 Angular velocity **87**; 8.3 Peripheral velocity **87**; 8.4 Power **88**; 8.5 Angular acceleration **89**; 8.6 Inertia of a rotating body **90**; 8.7 The flywheel **90**; Additional questions **91**

9 Expansion of Solids and Liquids

9.1 Linear expansion of a solid **92**; 9.2 Superficial expansion of a solid **93**; 9.3 Cubical expansion of a solid or a liquid **93**; 9.4 Practical applications **94**; Additional questions **96**

10 The Gas Laws

10.1 The effect of heat on a gas **97**; 10.2 Absolute unit of temperature **97**; 10.3 Absolute units of pressure **98**; 10.4 The gas laws **98**; 10.5 Heat as a form of energy **101**; 10.6 Heat transfer **104**; Additional questions **104**

11 An Introduction to Electricity

11.1 The gold leaf electroscope **106**; 11.2 The construction of materials **106**; 11.3 The process of electrical charging **108**; 11.4 Electrical conductors **109**; 11.5 Electrical insulators **109**; 11.6 Electric field **109**; 11.7 Electromotive force **109**; 11.8 Quantity of electric charge **110**; 11.9 Capacity **111**; 11.10 Electric current **111**; 11.11 Electrical resistance **112**; 11.12 Semiconductors **113**; 11.13 Heat energy released by an electric current **113**; Additional questions **114**

12 The Electrical Circuit

12.1 Electric circuit **115**; 12.2 Electric cell **115**; 12.3 The measurement of current **116**; 12.4 The measurement of potential difference **116**; 12.5 Ohm's law **116**; 12.6 Power developed by a resistance **117**; 12.7 Resistors **118**; 12.8 The series circuit **118**; 12.9 The parallel circuit **120**; 12.10 The internal resistance of cells **123**; 12.11 Cells connected in series **124**; 12.12 Cells connected in parallel **126**; 12.13 A series–parallel circuit **126**; Additional questions **129**

13 Resistivity and the Effect of Temperature

13.1 Resistivity **131**; 13.2 Conductance **135**; 13.3 Conductivity **136**; 13.4 Effect of temperature change on the value of resistance **137**; 13.5 The effect of positive temperature coefficient **138**; 13.6 Alloyed resistance wire **139**; 13.7 The effect of negative temperature coefficient **140**; 13.8 Semiconductor material **140**; Additional questions **141**

14 The Use of Ammeters and Voltmeters

14.1 Ammeter–voltmeter range extension **143**; 14.2 The effect on a circuit of the application of a voltmeter **145**; 14.3 A simple ohmmeter **149**; 14.4 The multi-range meter **150**; Additional questions **151**

15 The Chemical Effect of an Electric Current

15.1 Electrolysis **154**; 15.2 Faraday's laws of electrolysis **155**; 15.3 Practical use of electrolytic action **156**; 15.4 Electroplating **162**; 15.5 Electrolytic corrosion **164**; Additional questions **166**

16 Magnetism

16.1 The inverse square law of attraction and repulsion **167**; 16.2 Field patterns **168**; 16.3 Definition of magnetic field **169**; 16.4 Magnetizing a piece of steel using other magnets **170**; 16.5 Magnetic circuit **171**; Additional questions **171**

17 Electromagnetism

17.1 Field produced by two parallel conductors **174**; 17.2 Field from a solenoid **175**; 17.3 Magnetomotive force **175**; 17.4 The 'Ohm's law' of the magnetic circuit **175**; 17.5 Absolute permeability **177**; 17.6 Types of magnetic materials **179**; 17.7 Force on a conductor in a magnetic field **179**; 17.8 Definition of the ampere **181**; Additional questions **181**

18 Electromagnetic Induction

18.1 The generation of an electromotive force by a permanent magnet and coil **183**; 18.2 Mutual induction **188**; 18.3 Self-induction **188**; 18.4 Lenz's law **189**; 18.5 Practical applications of induced electromotive force **190**; Additional questions **191**

19 An Introduction to Alternating Current, the Generator, and Motor Effect

19.1 The rotation of a wire loop in a constant, parallel magnetic field **193**; 19.2 The motor effect **200**; Additional questions **201**

Answers to Additional Questions 203

Index 207

Preface

This book covers the engineering science and principles required for the first year of the City and Guilds of London Institute courses for telecommunication technicians (No. 271), electrical technicians (No. 281), electrical installation technicians (No. 285), and electronics technicians (No. 272). It also covers the general engineering course (No. 250).

A large number of worked examples have been included. In most cases a similar question follows a worked example so that the student may try for himself and more clearly understand the principle involved.

The additional questions at the end of each chapter range over the whole chapter material. Many of the questions are taken from, or are based upon, past City and Guilds examination papers and I am grateful to the Institute for permission to reproduce them. The accuracy of the answers is, of course, my own responsibility.

The units used in the text are those of the International System (SI) and practical derivations of them. Extracts from British Standards 3763 and 1991 are reproduced by permission of the British Standards Institution, 2 Park Street, London W1Y 4AA, from whom copies of the complete publications may be obtained.

The electrical symbols follow B.S. 3939, the first choice of symbol being used. Electrical circuits have been drawn to make their function clear. The potential fall is in most cases vertical as I believe that this makes the circuit easier to follow.

I am grateful to my friends for suggestions and would particularly like to thank Mr K. Tempest, C.ENG., MIERE, MSERT, for his help in the origination of this book.

T.A.L.

To my wife, Joan

List of Symbols and SI Units

Quantity	Symbol	Unit	Name of unit	Multiple and submultiple
acceleration	a	m/s²		
amplification factor	μ	–		
angle (plane)	$\alpha, \beta, \ldots$	rad	radian	m
apparent power	S	VA		
capacitance	C	F	farad	μ, p
conductance	G	S	siemens	
conductivity	σ	S/m		
density	ρ	kg/m³		
efficiency	η	–		
electric charge or flux	Q	C	coulomb	k, μ, n, p
electric current	I	A	ampere	k, m, μ, n, p
electric flux density	D	C/m²		
electric field intensity	E	V/m		
electric potential (electromotive force)	E, V	V	volt	
energy or work	W	J	joule	G, M, k, m
force	F	N	newton	M, k, m, μ
frequency	f	Hz	hertz	G, M, k
friction, coefficient	μ	–		
heat, latent	L	J	joule	M, k, m
heat, quantity	Q	J	joule	M, k, m
heat, total	H	J	joule	M, k, m
impedance	Z	Ω	ohm	
inductance, self	L	H	henry	
inductance, mutual	M	H	henry	
length	$l,$	m	metre	k, c, m, μ, n
loss angle	δ	rad	radian	
magnetic flux	Φ	Wb	weber	m
magnetic flux density	B	T	tesla	m
magnetic field intensity	H	At/m, A/m		
magnetomotive force	F	At, A	ampere-turn	
mass	m	kg	kilogramme	Mg, g mg, μg 1 tonne = 1 Mg
modulus of elasticity (Young's)	E	N/m²		

Quantity	Symbol	Unit	Name of unit	Multiple and submultiple
momentum	P	kg m/s		
number of turns	N	–		
period	τ	s	second	
permeability, absolute	μ	H/m		
permeability, free space	μ_0	H/m		
permeability, relative	μ_r	–		
permittivity, absolute	ε	F/m		
permittivity, free space	ε_0	F/m		
permittivity, relative	ε_r	–		
phase angle	ϕ	rad	radian	
power	P	W	watt	G, M, k, m, μ, n, p
pressure and stress	P	Pa	pascal	
reactance	X	Ω	ohm	
reactive volt amp	Q	var		
reluctance	S	At/Wb or 1/H		
resistance	R	Ω	ohm	
resistivity	ρ	Ωm		
rotational frequency	n	1/s or rev/s	reciprocal second	
temperature, absolute	T	K	kelvin	
temperature ('customary')	θ	°C	degree Celsius	
temperature (interval)	θ	°C or K		
time	t	s	second	k, m, μ, n
time constant	τ	s	second	
torque	M or T	N m		
velocity, linear	u or v	m/s		
velocity, angular	ω	rad/s		
weight	G	N	newton	

1 The International System of Units

Throughout history there has been an ever-increasing need to adopt a standardized method of expressing units since many of the various branches of science tended to use their own systems, causing difficulties in general understanding. Now that engineering has become more scientific, it must also use scientific units.

Scientists found the most useful method of expressing units to be a metric system which evolved after the French revolution; this was the centimetre-gramme-second (CGS) system. Early in the twentieth century the metre-kilogramme-second (MKS) system was developed, and in 1935 the ampere was added, thus forming the metre-kilogramme-second-ampere (MKSA or Giorgi) system.

In 1960 an international committee decided to add the degree Kelvin and the candela, making the Système International d'Unités or SI units.

The SI system is called a coherent system; that is, as the unit of length is 1 metre, the unit of area will be $1 \times 1 = 1$ square metre (1 m^2). The simplicity of the system will be more obvious if we compare, for example, the British acre, which is 4840 square yards.

1.1 Basic units

The six basic units of the SI system are:

Table 1.1

Property	Unit	Abbreviation	Symbol*
Length	metre	m	L or l
Mass	kilogramme	kg	M or m
Time	second	s	T or t
Electric current	ampere	A	I or i
Absolute temperature	degree Kelvin	K	T or t†
Luminous intensity	candela	cd	I or i‡

* As an alternative, upper-case characters may be used if the meaning is clear. The characters are printed in an *italic* (sloping) style.
† If likely to be confused with time then theta, Θ or θ, may be used.
‡ If it is likely to be confused with electric current, J or j may be used.

The following multiples and submultiples are used with SI units:

Table 1.2
Common multiples and submultiples

Name	Abbreviation	Power of 10	Name	Abbreviation	Power of 10
tera	T	10^{12}	centi	c	10^{-2}
giga	G	10^{9}	milli	m	10^{-3}
mega	M	10^{6}	micro	μ	10^{-6}
kilo	k	10^{3}	nano	n	10^{-9}
hecto	h	10^{2}	pico	p	10^{-12}

Examples of use: (a) 8000 m = 8×10^{3} m, which would be written as 8 kilometres or 8 km; (b) 0·003 m = 3×10^{-3} m, which would be written as 3 millimetres or 3 mm.

1.2 Derived units

All SI units are derived from the six basic units listed in Table 1.1. Starting from the units of length (l) and time (t)—where length is given in metres (m) and time in seconds (s)—the unit of velocity may be developed.

Velocity (u or v)

If a body* covers a distance (length) in a certain time then the body is said to be moving at a speed. If the direction of the movement is also specified then the body is said to have **velocity**.

Velocity may therefore be defined as a rate of change of displacement in a particular direction.

$$\text{Velocity} = \frac{\text{Change of length}}{\text{Time taken for this change}}$$

$$v = \frac{l_2 - l_1}{t_2 - t_1}$$

The symbol Δ (delta) can be used as a form of shorthand and velocity may then be stated as

$$v = \frac{\Delta l}{\Delta t}$$

The units of velocity are those of length divided by time; that is, metres/second or m/s. When speaking this is referred to as 'metres per second'.

Example 1.1 A man leaves his work at 17.00 hours and arrives home

* The term 'body' is used in mechanics to denote any object that is being discussed.

at 17.20 hours. If the distance between his work and his home is 8 km, what is his average velocity for the journey?

$$\text{Length} = 8 \times 10^3 \text{ m}$$
$$\text{Time taken, } t = 20 \times 60 = 1200 \text{ s} = 1{\cdot}2 \times 10^3 \text{ s}$$
$$\text{Average velocity} = \frac{\Delta l}{\Delta t} = \frac{8 \times 10^3}{1{\cdot}2 \times 10^3} = 6{\cdot}67 \text{ m/s}$$

Question 1.1 A car is moving at a velocity of 20 m/s. How long will it take to travel 1 km? (50 s)

Acceleration (a)

Acceleration is the rate of increase of velocity. For example, if a car is travelling at a velocity of 13 m/s and it takes 10 s to increase its velocity to 28 m/s, then its average acceleration would be *the change of velocity divided by the time taken for that change*, that is,

$$\frac{28 - 13}{10} = \frac{15}{10} = 1{\cdot}5 \text{ m/s}^2$$

Therefore acceleration is

$$\frac{\text{Change of velocity}}{\text{Time for that change}} \quad \text{or} \quad \frac{\Delta v}{\Delta t}$$

The units of acceleration shown above are metres per second squared (m/s²), sometimes said as 'metres per second per second' and these can be seen to be obtained from the units of velocity and time:

$$\text{The units of } \frac{\Delta v}{\Delta t} \text{ are } \frac{\text{m/s}}{\text{s}}$$

which can be written as m/s/s or m/s².

Deceleration

Deceleration is the rate at which velocity is reduced; it can also be considered as a negative acceleration.

Example 1.2 From a standing start a motor cycle can accelerate to 25 m/s in a time of 5 s. Find the average acceleration.

$$\text{Average acceleration} = \frac{\Delta v}{\Delta t} = \frac{25}{5} = 5 \text{ m/s}^2$$

Question 1.2 If the motor cycle of Example 1.2 takes 10 s to stop from a velocity of 25 m/s, find its average deceleration. (2·5 m/s²)

1.3 Distance–time graph

A graph is a very useful method by which, for example, progress can be immediately observed.

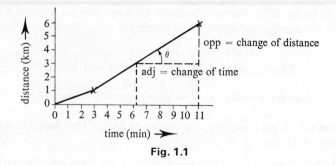

Fig. 1.1

If a car travels 1 km in a time of 3 min then a further 5 km in a time of 8 min, this could be shown by drawing a distance–time graph (Fig. 1.1).

The graph line will show when the car is travelling at the higher speed; the steeper the gradient on the graph, the higher the speed.

The velocity or speed may be calculated from this graph by finding the gradient of the graph. The method is to divide the opposite side to angle θ (opp.) by the adjacent side to angle θ (adj.). This ratio, opp./adj., is the same as

$$\frac{\text{Change of distance}}{\text{Time for this change}} \quad \text{or} \quad \frac{\Delta l}{\Delta t}$$

which is velocity.

From the triangle constructed in Fig. 1.1 the opposite side measures 3 km on the distance scale and the adjacent side 4·8 min on the time scale. Thus

$$\text{Velocity} = \frac{\text{opp.}}{\text{adj.}} = \frac{3 \times 1000 \text{ (metres)}}{4 \cdot 8 \times 60 \text{ (seconds)}}$$

$$= 10 \cdot 4 \text{ m/s average speed}$$

Question 1.3 A car covers a distance of 5 km in 10 min then slows and covers the next 2 km in 6 min. Draw a distance–time graph and find the average speed of the car for the first 10 min and for the following 6 min.
(8·33 m/s; 5·56 m/s)

1.4 Velocity–time graph

The velocity–time graph is more useful than the distance–time graph as both distance and acceleration may be obtained from it. The graph line A–B in Fig. 1.2 shows a condition of constant acceleration from zero as the line has a constant gradient and is sloping upwards.

The gradient is opp./adj., which to the graph scales is

$$\frac{\text{Change of velocity}}{\text{Time for this change}} \quad \text{or} \quad \frac{\Delta v}{\Delta t}$$

which is acceleration.

4 THE INTERNATIONAL SYSTEM OF UNITS

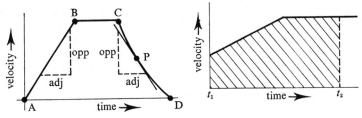

opp = change of velocity
adj = change of time   **Fig. 1.2**   **Fig. 1.3**

Section B–C shows the condition of constant velocity. The acceleration can be seen to be zero as the line has no gradient.

Section C–D shows a condition of variable deceleration as the line curves in a downward direction. The gradient at any particular point, for example P, may be found by drawing a tangent to the graph at that point and finding the gradient of the tangent.

The distance covered by the body may be determined by finding the scale area enclosed by the graph.

Thus the hatched area of the graph in Fig. 1.3 will, to the graph scales, be the distance covered between t_1 and t_2. Since the velocity equals the distance covered divided by the time taken, or $\Delta l/\Delta t$,

$$\text{Distance covered} = \text{Velocity} \times \text{Time taken}$$

or the scale area of the graph.

Example 1.3   A car accelerates at a constant rate from zero to 20 m/s in a time of 5 s (Fig. 1.4). Find (a) the acceleration and (b) the distance covered.

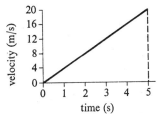

Fig. 1.4

(a) Acceleration = Gradient to the graph scales

$$= \frac{\text{opp.}}{\text{adj.}} = \frac{20 \text{ (m/s)}}{5 \text{ (s)}} = 4 \text{ m/s}^2$$

(b) Distance   = Area to the graph scales
  = $\frac{1}{2}$ base × height
  = $\frac{1}{2} \times 5 \times 20 = 50$ m

Question 1.4   If the car in Example 1.3 accelerates from 20 m/s to 30 m/s in a further time of 5 s, sketch the velocity–time graph and find (a) its acceleration and (b) the further distance covered.   (2 m/s²; 125 m)

Example 1.4 A ball, thrown vertically upwards, returns to the thrower in 2 s (Fig. 1.5). Find (a) the initial velocity and (b) the height attained by the ball. The acceleration due to gravity is 9·8 m/s².

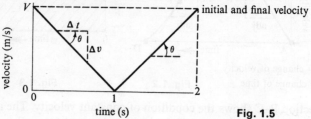

Fig. 1.5

The gradient of the graph will be 9·8 to the graph scales, and if the initial velocity is V, the velocity–time graph can be sketched.

As the ball moves upwards its velocity will fall from V m/s to zero, and as it returns its velocity will increase from zero to V m/s. The time taken for the ball to reach its maximum height will be one-half of the total time, that is 1 s. The deceleration and acceleration will both be 9·8 m/s². Therefore, from gradient = $a = \Delta v/\Delta t$,

$$\Delta V = 9\cdot 8 \times 1 = 9\cdot 8 \text{ m/s}$$

Therefore, (a) the initial velocity will be

$$9\cdot 8 - 0 = 9\cdot 8 \text{ m/s.}$$

The height attained will be the distance travelled when either going up or coming down. Therefore, (b) the height will be

$$\text{Area of either half of the graph} = \tfrac{1}{2} \times 9\cdot 8 \times 1$$
$$= 4\cdot 9 \text{ m}$$

An allowance must also be made for the height of the thrower, so that the height above ground is 4·9 *plus* height at the release of the ball.

Question 1.5 A ball which is thrown vertically upwards takes 3 s to return to the thrower. Find (a) its initial velocity and (b) the total distance travelled by the ball. The acceleration due to gravity is 9·8 m/s².   (14·7 m/s; 22·05 m)

1.5 Mass (*m*)

The **mass** of a body is the amount of material contained by the body.

Density

The definition of mass contains no reference to size. For example, a small piece of lead has a relatively large mass and is therefore said to have a high **density**. On the other hand, a piece of aluminium of the same mass will have a much larger volume, and its density is said to be less than that of lead.

The density of a body is the mass for a certain volume and is measured in kilogrammes per cubic centimetre (kg/cm³).

Relative density

By definition, 1 cm³ of pure water has a mass of 1 gramme (g) and thus its density is 1 g/cm³. **Relative density** is the density of a body related to that of water. If the relative density is greater than 1 the object will sink in water, but if it is less than 1 it will float.

$$\text{Relative density} = \frac{\text{Mass of the body}}{\text{Mass of an equal volume of water}}$$

Example 1.5 A piece of aluminium has a mass of 25·6 g and a volume of 10 cm³. Find its relative density and its density in grammes per cubic centimetre and in kilogrammes per cubic centimetre.

10 cm³ of aluminium has a mass of 25·6 g

10 cm³ of water has a mass of 10 g

therefore,

$$\text{Relative density} = \frac{\text{Mass of the aluminium}}{\text{Mass of an equal volume of water}}$$

$$= \frac{25 \cdot 6}{10} = 2 \cdot 56$$

and

$$\text{Density} = \frac{\text{Mass}}{\text{Volume}} = \frac{25 \cdot 6}{10} = 2 \cdot 56 \text{ g/cm}^3$$

or

$$= \frac{25 \cdot 6/1000}{10} = 0 \cdot 002\ 56 \text{ kg/cm}^3$$

From these calculations it can be seen that the relative density is the same as the density if the units are grammes per cubic centimetre.

A name sometimes used for relative density is **specific gravity** and its use may still be noticed in some tables.

Examples of relative density

Table 1.3

Material	Relative density	Material	Relative density
Aluminium	2·56	Mercury	13·6
Chromium	6·93	Nickel	8·8
Copper	8·65	Tin	7·3
Lead	11·4	Zinc	7·0

An interesting example of the practical use of varying relative density is the submarine. If the mass of the submarine is greater than a mass of an equal volume of water then it will sink, and if the mass is less it will rise.

Water is expelled from tanks within the submarine to increase its volume of air, and hence its buoyancy, when it wishes to surface.

Question 1.6 By the use of Table 1.3 find the mass, in kilogrammes, of 1 m³ of lead. (11 400 kg)

Additional questions

1.7 A stone, dropped from a tower, is timed to take 3 s to reach the ground. If it falls at the acceleration of gravity (9·8 m/s²) sketch the velocity-time graph and find (a) the height of tower and (b) the velocity at which it strikes the ground.

1.8 A lift is constantly accelerated from rest for a time of 5 s. During this time the distance covered is 5 m. Find its final velocity and the value of the acceleration.

1.9 If the lift of Question 1.8 covers a further 10 m without any increase in velocity, find the time taken.

1.10 A stone is thrown directly downwards with an initial velocity of 5 m/s. If the stone takes 4 s to reach the ground, find (a) the height from which it was thrown and (b) its velocity on reaching the ground. Take the acceleration due to gravity as 9·8 m/s².

1.11 A car starting from rest, accelerates in a uniform fashion to 13·4 m/s in a time of 6 s, and immediately accelerates in a uniform fashion to 22·3 m/s in a further 9·6 s. Find the distance travelled after (a) 6 s and (b) 15·6 s.

1.12 Find the uniform acceleration of the car of Question 1.11 during (a) the first 6 s and (b) the second 9·6 s.

1.13 A solid piece of lead has a mass of 1 kg. Find its volume if the relative density of lead is 11·4.

1.14 A solid block of copper has dimensions of 8 cm long, 4 cm high, and 3 cm wide. Find its mass in kilogrammes if the relative density of copper is 8·65.

1.15 A container has a total volume of 1 litre (1000 cm³) and is half filled with pure water. A solid piece of metal having a mass of 640 g is inserted into the container which is then found to be three-quarters filled. Find the relative density of the metal and estimate from your answer and the table of relative densities (Table 1.3) the type of metal.

2 Newton's Laws of Motion

Sir Isaac Newton, who was born in 1642 in the reign of Charles I, formulated three laws of motion. These may be referred to, respectively, as the laws of inertia, momentum, and reaction.

2.1 The law of inertia

Unless a body is acted upon by an external force, it will continue in its existing state of rest, or of motion, in a straight line at constant velocity.

This law may be used to formulate a definition of a force.

Force

Any push or pull which changes the existing state of rest, motion, or direction of a body is called a **force**.

The manual movement of a motor car is an excellent example by which to demonstrate the law of **inertia**. A great effort is required to start the movement, but once moving on level ground the force may be reduced. Again, when stopping, a large reverse force must be applied. If the direction of the car is altered while the car is in motion, the force must be increased.

This inertia effect of change of direction is used, for example, by ships when they have to stop quickly in an emergency. By steering a zig-zag course the inertia of the change of direction helps to balance the inertia trying to continue the ship's movement.

2.2 The law of momentum

If a force has the ability to move a body, then the rate of change of momentum will be proportional to the force and the direction of the movement will be that of the force.

Momentum (P)

$$\text{Momentum} = \text{Mass} \times \text{Velocity}$$
$$P = m \times v$$

In SI units this is written as

$$\text{kg m/s} = \text{kg} \times \text{m/s}$$

Thus the SI units of **momentum** will be kg m/s.

Example 2.1 A mass of 10 kg has a velocity of 8 m/s. Find its momentum.
$$P = m \times v = 10 \times 8 = 80 \text{ kg m/s}$$

Question 2.1 A mass of 50 g has a momentum of 0·25 kg m/s. Find its velocity.
(5 m/s)

The relationship between mass and force

Let a force act on a body of mass m kilogrammes and let its velocity be increased from v_1 to v_2 metres per second in a time of t seconds.

The average rate of change of momentum will be:

$$\frac{mv_2 - mv_1}{t} \quad \text{or} \quad \frac{m \, \Delta v}{\Delta t}$$

where $\Delta v / \Delta t$ is the acceleration. Therefore

Average rate of change of momentum $= ma$

From the second law of motion the force is proportional to the rate of change of momentum:

$$F \propto ma.$$

In the SI system the unit of force is chosen to make this formula

$$F = ma$$

The SI unit is appropriately called the **newton (N)** and by use of the formula $F = ma$ the newton may be defined.

The unit of force—the newton

The newton is that force which, when applied to a body having a mass of one kilogramme, gives an acceleration of one metre per second squared (B.S. 3763).

$$F(\text{N}) = m(\text{kg}) \times a(\text{m/s}^2)$$

Example 2.2 A body of mass 10 kg is to be given an acceleration of 5 m/s². How much force must be applied?

$$F = ma$$

We know that $m = 10$ kg and $a = 5$ m/s², therefore

$$F = 10 \times 5 = 50 \text{ N}$$

If the units are not the absolute SI units they must first be converted.

Example 2.3 A mass of 250 g is to be given an acceleration of 50 cm/s². Find the force required.

$$\text{Mass} = \frac{250}{1000} = 0.25 \text{ kg}$$

$$\text{Acceleration} = \frac{50}{100} = 0.5 \text{ m/s}^2$$

therefore,
$$F = Ma = 0.25 \times 0.5 = 0.125 \text{ N}$$

Question 2.2 A mass of 2 kg has a force of 10 N acting on it. Find the acceleration of the mass. (5 m/s^2)

Question 2.3 A mass of 850 g has a force of 15 N acting on it. Find the acceleration of the mass in (a) metres per second squared, and (b) centimetres per second squared. (17.65 m/s^2; 1765 cm/s^2)

Weight (G)

Weight is the natural downward force due to the mass of an object. By this definition, weight is a force and the SI unit of measurement will be the newton.

When the acceleration of gravity is acting on a mass of 1 kg the downward force (weight) will be:

$$G = m \times g = 1 \text{ (kg)} \times 9.81 \text{ (m/s}^2\text{)} = 9.81 \text{ N}$$

In everyday use, mass is loosely spoken of in terms of weight. For example, if one buys a 'kilo' (kilogramme) of apples the greengrocer 'weighs' out the apples, implying the purchase of a weight of apples. The purchase made is in fact a *mass* of a kilogramme of apples.

A *non*-SI unit of weight which may be used in practical contexts is the **kilogramme-force**. The kilogramme-force (kgf) is the weight of a mass of 1 kilogramme when the acceleration acting on it is that of gravity, that is 9.81 m/s^2.

From this definition of a kilogramme-force it will be seen that the kilogramme-force is the equivalent of 9.81 N.

The weight of a mass of 1 kg will only be 1 kgf (9.81 N) when the acceleration of gravity is 9.81 m/s^2.

To prove that the weight of an object is not an absolute constant, take a spring balance and a mass of 1 kg into a lift. The spring balance will indicate a force of 1 kgf (9·81 N) when the lift is stationary or moving with constant velocity. As the lift accelerates upwards with an acceleration of a m/s² the acceleration acting on the kilogramme mass is $(g + a)$ m/s² and the weight will increase to:

$$G = m(g + a) = 1(9.81 + a) \text{ N}$$

In the kilogramme-force unit this will be:

$$\text{Weight} = \frac{9.81 + a}{9.81} \text{ kgf}$$

When the lift decelerates going upward the acceleration falls to $(g - a)$ m/s² and the weight will reduce. It is interesting to note the change in your body weight when in a lift although it is obvious that your body mass cannot change.

Question 2.4 When travelling downwards in a lift when will a person feel (a) lighter and (b) heavier? (*Answer given at end of book*)

The measurement of mass and force

From the previous example it may be concluded that a spring balance measures force. The chemical balance shown in Fig. 2.1 will balance mass since gravity will effect both scale pans equally.

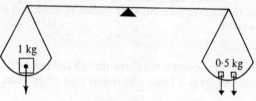

Fig. 2.1

On the left-hand scale pan

$$G = mg = 1 \times 9.81 = 9.81 \text{ N}$$

On the right-hand scale pan

$$G = m_1 g + m_2 g = (0.5 \times 9.81) + (0.5 \times 9.81) = 9.81 \text{ N}$$

Thus the two forces cause no movement.

2.3 The reaction force law

If a body is stationary or moving with constant velocity then any applied force must have an equal and opposite force (Fig. 2.2).

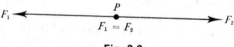

Fig. 2.2

Either of these forces may be a combination of forces. For example, the reaction force may consist of a friction force and the force of air upon the body.

Example 2.4 Two opposing tug-of-war teams pull with an equal force of 3000 N on the ends of a rope so that the rope remains stationary (in equilibrium). Find the force in the rope.

The force in the rope will be 3000 N as one team acts as the reaction force to the other. If one team had tied its end of the rope to a strong tree, the force in the rope would still be 3000 N.

Question 2.5 If one end of the rope in Example 2.4 is tied to a firm anchorage and both teams pull on the same end of the rope with an effort of 3000 N, find the force in the rope. (*Answer given at end of book*)

2.4 Friction force

When a body is to be moved from rest a certain minimum force will be required before the body will start to move. This force is said to be the force required to overcome **static friction.** When the body begins to move the minimum force required to keep it moving with constant velocity is known as the force required to overcome **sliding friction.** The sliding friction force is always less than the static friction force.

If the surface between the body and its support were examined under a microscope it would have an appearance similar to that shown in Fig. 2.3.

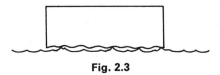

Fig. 2.3

Before it can be moved the body must be lifted over the larger 'hills' but once over the hill it will slide down the other side thus assisting its progress up the next hill.

In addition, the force to overcome static friction must include an accelerating force component to provide even the smallest movement.

The reaction force due to friction is proportional to the force between the friction surfaces and to the type of surface.

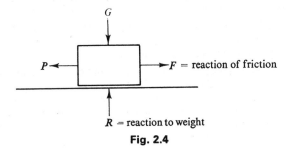

Fig. 2.4

Coefficient of friction (μ)

For a given type of surface a constant can be obtained by dividing the minimum force required for constant velocity by the force holding the surfaces together.

$$\text{Coefficient of friction, } \mu = \frac{P}{G}$$

The friction force reaction will be opposite and equal to the force P, and there will also be an upward reaction to the weight, force G (Fig. 2.4). Thus

$$\mu = \frac{P}{G} \quad \text{or} \quad \frac{F}{R}$$

Example 2.5 A mass of 2 kg stands on a horizontal surface and it is found that the smallest force that can make it start to slide is 3 N. Find the coefficient of static friction for the surfaces.

$$\text{Downward force } G = ma = 2 \times 9\cdot81 = 19\cdot62 \text{ N}$$

$$\mu = \frac{P}{G} = \frac{3}{19\cdot62} = 0\cdot153$$

Question 2.6 A mass of 3 kg stands on a horizontal surface. If the coefficient of friction between the surfaces is 0·35, find the force required (a) in newtons and (b) in kilogrammes-force, to overcome friction. (10·3 N; 1·05 kgf)

Additional questions

2.7 Find the force in newtons and in kilogrammes-force required to accelerate a mass of 2 kg at a rate of 3·44 m/s².

2.8 A mass of 30 g is given an acceleration of twice that of gravity. Find the force required (a) in newtons and (b) in grammes-force.

2.9 A mass of 2 kg is moving in a straight line with a velocity of 10 m/s. Find its momentum (a) in kilogramme-metres per second and (b) in gramme-centimetres per second.

2.10 Calculate the average force required in newtons to accelerate a 5 kg mass from 5 m/s to 15 m/s in a time of 2 s.

2.11 A mass of 2 kg is weighed in a descending lift and found to be 2·3 kgf. State whether the lift is accelerating or decelerating and find the value of acceleration or deceleration.

2.12 A mass of 5 kg is acted on by a force of 60 N. If the coefficient of friction between the mass and the horizontal surface on which it rests is 0·7, find the resulting acceleration of the mass.

2.13 Explain what is meant by the term 'force', and define two units of force.
A body of mass 25 kg, originally at rest, is given a constant acceleration of 2 m/s² for 20 s, then an acceleration of 4 m/s² in the same direction for a further 20 s. Calculate (a) the velocity at the end of 40 s, (b) the force, in newtons, required during the first 20 s, and (c) the force, in newtons, required during the last 20 s. (Based upon C.G.L.I.)

2.14 (a) Distinguish between a mass of 4 kg and a weight of 4 kgf. Taking the acceleration due to gravity as 9·81 m/s², find the gravitational force on a 10 kg mass in newtons.
(b) A vertical rope is used to give an upward acceleration of 5 m/s² to a 50 kg mass attached to its lower end. What is the tension in the rope? If the maximum safe tension in the rope is 100 kgf, what is the corresponding maximum acceleration of the mass?
(Based upon C.G.L.I.)

3 The Effects of Force

3.1 Stress

In practice, a structural member must have sufficient cross-sectional area to resist an applied force. The applied force is distributed over this cross-section and it is said that a structural member having a small cross-sectional area is more highly stressed than one with a large cross-sectional area.

An example of stress is to consider a roof supported (as shown in Fig. 3.1) by thin pillars and the same roof supported (as shown in Fig. 3.2) by thick pillars. The pillars shown in Fig. 3.2 are less stressed

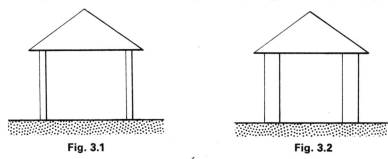

Fig. 3.1　　　　　　　　　**Fig. 3.2**

than those shown in Fig. 3.1, and if they are made of a similar material they are less likely to bend and break.

From these examples it is clear that it is important to consider both the load and the cross-sectional area in one term, which is called **stress**. Thus

$$\text{Stress} = \frac{\text{Load}}{\text{Cross-sectional area}}$$

The SI unit of stress is the Pascal (Pa) which is 1 N/m^2.

Example 3.1 A wire has a load of 2 kN applied to it. Find the stress if the wire has a diameter of 2 mm.

$$\text{Cross-sectional area of wire} = \frac{\pi d^2}{4} = \frac{\pi \times (2 \times 10^{-3})^2}{4}$$

$$= 3\cdot142 \times 10^{-6} m^2$$

$$\text{Stress} = \frac{\text{Load}}{\text{Area}} = \frac{2 \times 10^3}{3\cdot142 \times 10^{-6}} = 636\cdot5 \text{ MPa}$$

Question 3.1 A pillar has a load acting down on it of 30 kN. The pillar is of square section having sides of 0·5 m. Find the stress, in pascals, in the pillar.
(120 kPa)

STRESS 15

3.2 Types of stress

Tension

If external forces are attempting to pull a structural member apart the member is said to be in **tension** (Fig. 3.3). The load or stress is referred to as a *tensile* load or stress.

F_A ←———○——— F_R wire F_R ———○——→ F_A

F_A = applied external forces
F_R = internal reactions of wire   **Fig. 3.3**

It is common practice to provide a cable or wire to resist this type of load. Practical examples of their use are: suspension wires on a suspension bridge; stay wires on telephone poles; tie wires on cranes.

A more complete answer to Example 3.1 would be 636·5 MPa *tension* as the load is a tensile load.

Compression

If the structural member is under a crushing load then it is said to be in **compression** (Fig. 3.4). The load or stress is said to be a *compressive* load or stress.

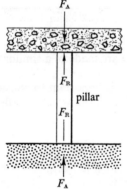

F_A = external applied forces to the pillar by the concrete beam at the top and the reaction of the ground below
F_R = internal reaction force within the pillar

Fig. 3.4

To withstand a compressive load a structural member must be rigid. Practical examples of their use are: struts on telephone poles; the jib of a crane; the piers of a bridge.

Shear

A structural member which is under a cutting or scissors-like load is under a **shear** load. The cross-sectional area of the rivet shown in Fig. 3.5 is the area resisting the shear. As this area only resists the shear once this is called a condition of single shear.

A condition of double shear is shown in Fig. 3.6, in which the rivet cross-sectional area is resisting the shear twice.

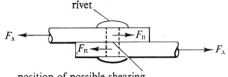
F_A = external applied forces
F_R = reaction force of the rivet to shear
Fig. 3.5

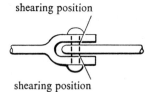
Fig. 3.6

Practical examples of a shearing load are: steering joints on cars; vehicle spring fixing bolts; metal cutting shears and guillotines.

Example 3.2 The load applied to the double shear joint shown in Fig. 3.6 is 1000 kgf and the rivet has a diameter of 5 mm. Find the shear stress.

Referring to Fig. 3.6, the area presented to shear is double the cross-sectional area of the rivet.

$$\text{Area of rivet} = \frac{\pi d^2}{4} = \frac{\pi \times (0.5)^2}{4} \text{ cm}^2$$

$$= \frac{\pi \times 0.25}{4} \text{ cm}^2$$

$$= 0.196 \text{ cm}^2$$

$$\text{Area presented to shear} = 2 \times 0.196 = 0.392 \text{ cm}^2$$

$$\text{Shear stress} = \frac{\text{Load}}{\text{Area}} = \frac{1000}{0.392} = 2560 \text{ kgf/cm}^2$$

Question 3.2 A rivet under single shear stress has a diameter of 8 mm. If the shear stress does not exceed 1000 kgf/cm², find the maximum permissible load. If the rivet was in double shear how would the maximum load be affected. (502 kgf; doubled)

3.3 Strength of materials

Table 3.1
Breaking or ultimate stress for materials with values of Young's modulus of elasticity

Material	Stress						Young's modulus E	
	Tension		Compression		Shear			
	kgf/cm²	MPa	kgf/cm²	MPa	kgf/cm²	MPa	kgf/cm²	Pa × 10⁹
Cast iron	1540	150·7	7000	686·7	1680	164·8	1 120 000	109·9
Carbon steel	5250	515	5250	515	3850	377·7	2 100 000	206
Nickel steel	9100	892·7	9100	892·7	6850	672	2 100 000	206
Aluminium	840	82·4	—	—	—	—	720 000	70·63
Brass	4200	412	—	—	—	—	1 050 000	103
Concrete	28	2·75	140	13·7	—	—	—	—

It is useful to note from Table 3.1 that in general a material is weakest when under a shear stress and this should be avoided whenever possible in structures.

3.4 Strain

When a material is loaded a physical distortion will take place as the load is resisted. If the load is increased above a maximum for the material, permanent distortion will take place which will eventually lead to destruction.

When a metal rod of length 1 metre is extended 1 μm (1 micron) it is more strained than a 10 m rod of the same metal also extended 1 μm.

Strain is therefore measured by considering both the change of length and the length before loading.

$$\text{Strain} = \frac{\text{Change of length}}{\text{Original length}}$$

$$\text{Tensile strain} = \frac{\text{Extension}}{\text{Original length}}$$

$$\text{Compressive strain} = \frac{\text{Reduction}}{\text{Original length}}$$

Example 3.3 A strut is under a compressive load. If the lengths before and after loading were 10 and 9·95 m respectively, find the strain.

$$\text{Compressive strain} = \frac{\text{Reduction}}{\text{Original length}} = \frac{10 - 9 \cdot 95}{10} = \frac{0 \cdot 05}{10}$$
$$= 0 \cdot 005$$

No units are used as strain is a ratio of two lengths.

Question 3.3 A wire of length 5 m has its length increased by 3 mm when a mass is suspended from it. Find the strain in the wire, and state the type of strain.   (0·0006; *second answer at end of book*)

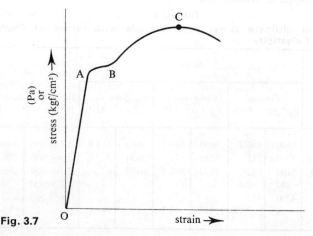

Fig. 3.7

18 THE EFFECTS OF FORCE

3.5 Relation between stress and strain

If a tensile test is carried out on a steel wire, and a graph is plotted to show the relation between stress and strain, the resulting shape will be similar to that shown in Fig. 3.7.

Initially the graph will be found to be a straight line, OA. During loading the material acts rather like a piece of rubber and this is known as the **elastic** stage. When the load is removed the material will return to its original length with no permanent distortion. Postition A on the graph is called the *elastic limit*, or limit of proportionality, for the material.

If this limit is exceeded the material becomes **plastic** and is easily altered in shape, and after position B the material rapidly passes to its **ultimate strength** position, C, after which it will break.

Hooke's law
Within the elastic limit, strain is proportional to stress. This is shown by the line OA in Fig. 3.7.

Example 3.4 A wire is stretched 1 mm by a load of 10 kgf. Find the load that would stretch the wire 3 mm assuming that the elastic limit is not exceeded.

This may be calculated using a proportion.

$$1 \text{ mm} : 100 \text{ N} :: 3 \text{ mm} : \text{required load}$$

$$\frac{1}{10} = \frac{3}{\text{Load}}$$

Therefore, the load must be

$$3 \times 10 = 300 \text{ N}$$

Question 3.4 If the load applied to the wire of Example 3.4 is 220 N, what extension will result? (2·20 mm)

Young's modulus of elasticity (E)
This is the calculated stress that would produce unit strain; that is, it is the estimated stress that would cause the material to double its length. This length can be estimated by projecting the OA line of Fig. 3.7 in the manner shown in Fig. 3.8.

The scales of the graph would need to be changed considerably as the stress required to produce a strain of 1 is extremely high. In practice this value of stress can never be reached without breaking the material, and its use is solely for calculation purposes.

From the definition of Young's modulus of elasticity: when the strain = 1, the stress = E, or

$$E = \frac{\text{Stress}}{1}$$

For any other value of stress

$$E = \frac{\text{Stress}}{\text{Strain}}$$

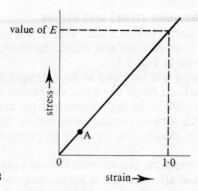

Fig. 3.8

By using the figures listed in Table 3.1 for Young's modulus of elasticity, and provided that the material is operated within its elastic limit, calculations of stress and strain may be made.

Example 3.5 A tensile load of 1000 kgf is applied to a wire of 0·5 cm diameter and of unloaded length 5 m. If the value of Young's modulus of elasticity is 2×10^6 kgf/cm², find the extension of the wire.

$$\text{Area of wire} = \frac{\pi d^2}{4} = \frac{\pi \times 0\cdot 5^2}{4} = 0\cdot 196 \text{ cm}^2$$

$$\text{Stress} = \frac{\text{Load}}{\text{Area}} = \frac{1000}{0\cdot 196} = 5100 \text{ kgf/cm}^2$$

But E = Stress/Strain, therefore Strain = Stress/E. Thus

$$\text{Strain} = \frac{5100}{2 \times 10^6} = 2\cdot 55 \times 10^{-3}$$

and since Strain = Extension/Original length,

$$\begin{aligned}\text{Extension} &= \text{Strain} \times \text{Original length} \\ &= 2\cdot 55 \times 10^{-3} \times 5 \\ &= 12\cdot 75 \times 10^{-3} \text{ m} \\ &= 12\cdot 75 \text{ mm}\end{aligned}$$

Question 3.5 A cable of 2·5 mm diameter has an unloaded length of 3 m. It is then subjected to a tensile load of 150 kgf. If the value of Young's modulus for the wire is 2×10^6 kgf/cm², find the increase in length at this load.

(4·578 mm)

Extension of a spring

Certain types of steel wire may be wound into a coil to form a coil spring. The spring can then be considerably extended without exceeding the elastic limit for the steel. The extension is obtained mainly by a bending rather than a stretching action.

The spring will obey Hooke's law; that is, the extension of the spring will be proportional to the load provided the elastic limit is not exceeded.

Example 3.6 A coil is stretched 3 cm by a load of 0·5 kgf. Find the extension caused by a load of 2 kgf assuming this load to be within the elastic limit.

$$\text{Extension} \propto \text{Force}$$

$$3 \text{ cm} : 0\cdot5 \text{ kgf} :: \text{Extension} : 2 \text{ kgf}$$

$$\text{Extension} = \frac{3}{0\cdot5} \times 2 = 12 \text{ cm}$$

This principle of extension, being proportional to force, is used in a spring balance where the free end of spring moves a pointer over a fixed scale.

Springs can also be wound into a spiral; for example, a watch mainspring and a balance-wheel spring. This type of spring is also in very common use in the moving-coil type ammeter or voltmeter.

Question 3.6 A spiral spring requires a force of 10 N to cause a deflection of 15°. If a force of 63 N is applied, what will be the angle of deflection?

(94·5°)

3.6 Factor of safety

A material must never be stressed to anywhere near its ultimate limit as unexpected loads may be encountered.

For example, a bridge may be expected to carry only ten cars at any one time. In an exceptional circumstance, say a traffic jam, the bridge might in fact have thirty cars standing on it. This is an overload of three times, or 300 per cent. In addition, there could be a layer of snow or a side wind.

In practice, therefore, a safety factor is allowed.

$$\text{Factor of safety} = \frac{\text{Ultimate strength}}{\text{Working stress}}$$

Example 3.7 A bridge notice states that the maximum allowed load is 5 Mgf, but it is known that in the design of the bridge a factor of safety of 5 was allowed. What is the ultimate load the bridge should be capable of withstanding?

$$\text{Factor of safety} = \frac{\text{Ultimate strength}}{\text{Working stress}}$$

Therefore

$$\text{Ultimate strength} = \text{Working stress} \times \text{Factor of safety}$$
$$= 5 \times 5 = 25 \text{ Mgf}$$

Question 3.7 A wire is to be stressed to 1500 kgf/cm². If a factor of safety of 4 is to be allowed find the ultimate strength required by the wire.

(6000 kgf/cm²)

Additional questions

3.8 Forces of 100 N are applied in opposite directions at either end of a cord 2 mm in diameter. Find the stress within the cord.

3.9 Two forces, each of 100 N, are applied in the same direction to the same end of a cord 2 mm diameter. If the other end of the cord is fixed to a wall, find the stress in the cord.

3.10 The strain on a wire is 0·001. If its length before loading was 5 m, find its increase in length.

3.11 A steel wire, 1 cm diameter and 3 m long, stretches 2·5 mm under a load of 10 kN. Find the modulus of elasticity.

3.12 A mass which is suspended from a steel wire 5 m long and 2 mm diameter causes the wire to stretch 3 mm. Find the size of the mass if the modulus of elasticity of the wire is 2 000 000 kgf/cm².

3.13 A carbon steel bolt of minimum diameter 1 cm is to be given a safety factor of 5. By using Table 3.1 find the maximum shear load allowable for the bolt.

3.14 What is meant by the terms 'tensile stress', 'tensile strain', and 'elastic limit'?

A tow rope has a circular section of 2 cm diameter. It is used to tow a vehicle of mass 1000 kg. The maximum acceleration of the vehicle during the tow is 1·5 m/s². Allowing for a frictional resistance of 140 N, what is the maximum stress in the rope? Explain why breakage of the tow rope is most likely to occur when 'taking up the slack'.

(C.G.L.I.)

3.15 In Fig. 3.9, link AB has vertically downward forces of 0·5 tonnef and 0·75 tonnef acting as shown, and it is supported by a steel rod CD. Find the stretch in rod CD given that it is 1 cm diameter, 2 m long, and that E for steel is 2×10^6 kgf/cm². (Based upon C.G.L.I.)

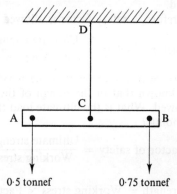

Fig. 3.9

4 The Pictorial Representation of Forces

4.1 Scalar quantities

A **scalar** quantity is one that has a magnitude but no direction. An example of a scalar quantity is mass. Mass is defined as *the quantity of material contained by a body*. This is merely a numerical value and irrespective of where the mass is positioned, this value will not change.

4.2 Vector quantities

Any quantity that has both magnitude and direction is termed a **vector** quantity. An example of a vector quantity is velocity. Velocity is a speed in a given direction and it follows, therefore, that speed is a scalar quantity and velocity is a vector quantity.

Pictorial representation of a vector quantity

Any type of vector quantity may be represented by drawing a line. A line will have length and position, and if the length of the line (to scale) represents the magnitude of the force, and its position represents the direction of the force, the line can be called the vector of the force.

Example 4.1 A rope is tied to a wall so that it makes an angle of 60° to the wall. A pull of 40 kgf is then exerted on the wall by the rope. Show this force by a vector diagram.

The first requirement is to construct a space diagram. This is a plan or elevation drawn to provide an accurate layout of the arrangement (Fig. 4.1). From this space diagram the vector diagram may be constructed.

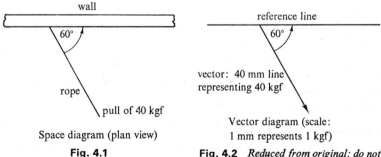

Fig. 4.1 Space diagram (plan view)

Fig. 4.2 Vector diagram (scale: 1 mm represents 1 kgf) *Reduced from original: do not scale*

Using the line of the wall as a reference, a vector is constructed representing the 40 kgf to scale.

The vector shown in Fig. 4.2 must be drawn parallel with the direction of rope shown in Fig. 4.1. The length of the vector will be 40 units, as 1 unit represents 1 kgf. Using the scale indicated in Fig. 4.2, 1 unit is 1 mm and so the vector will be 40 mm, or 4 cm in length.

The arrow drawn at the end of the vector indicates the direction of application of the force.

Example 4.2 A stay wire is attached to a telephone pole at a height of 8 m above ground level. The other end of the stay wire is anchored to the ground 4 m from the base of the pole. If the stay wire is tensioned to 200 kgf, draw a vector diagram representing the force acting on the pole.

A space and vector diagram may be constructed to a scale, as shown in Fig. 4.3.

Space diagram (elevation)
(Scale: 1 cm represents 2 m)

Vector diagram
(scale: 1 cm represents 40 kgf)

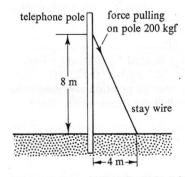
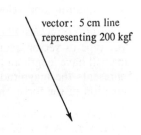

Fig. 4.3 *Reduced from original: do not scale*

Question 4.1 A ship is towed by a tug, the line of the cable making an angle, in the horizontal plane, of 10° with the centre-line of the ship. If the tension in the cable is 550 kgf, draw (a) a space diagram and (b) a vector diagram indicating the force acting on the ship.

4.3 The use of vector diagrams

The addition of forces in the same plane

If two forces are acting on the same point they can be replaced in effect by one force termed a 'resultant' force. To find this single force the two forces are added vectorially, that is, they are added taking both their size and their direction into consideration.

Example 4.3 Two forces of 5 and 7·5 N act on the same point as shown by Fig. 4.4. Find the single force in the same plane that could replace both.

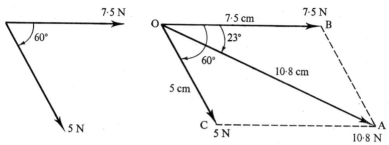

Fig. 4.4 *Reduced from original: do not scale*

The forces are first drawn as vectors OB and OC (Fig. 4.4).

A line parallel with OB is drawn from point C, and another line parallel with OC is drawn from point B. These two construction lines intersect at point A. The vector of the resultant force will be a line joining A and O. This line measures 10·8 cm, and from the vector scale this represents 10·8 N. The angle AOB measures 23° and the answer can now be expressed:

The resultant force is 10·8 N, acting at an angle of 23° from the 7·5 N force.

As the shape formed by the vectors and construction lines is a parallelogram, this diagram is called a 'parallelogram of forces'.

Question 4.2 The direction of action of the 5 N force in Example 4.3 is reversed, as shown by Fig. 4.5. Find the new value and direction of the resultant force. (6·7 N, 41° from the 7·5 N force)

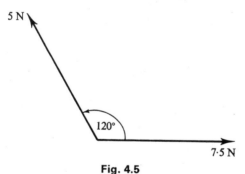

Fig. 4.5

If the point O in Fig. 4.4 is to remain in equilibrium, then a force opposite and equal to the resultant force must be applied. This force is termed an 'equilibrant' force and can be seen in Fig. 4.6.

If the equilibrant and resultant forces are vectorially added, the result will be zero.

Equilibrium must also result from the vectors in Fig. 4.7.

THE USE OF VECTOR DIAGRAMS   **25**

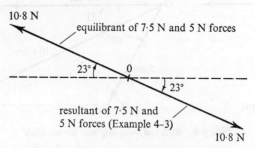

Fig. 4.6 *Reduced from original: do not scale*

4.4 The theorem of the triangle of forces

The vectors of Fig. 4.7 can construct the triangle shown in Fig. 4.8.

The theorem of the triangle of forces states that if three forces are (a) in equilibrium, (b) not parallel, and (c) acting in the same plane, then their vectors will form a triangle, and the direction of action of the vectors will follow around the triangle.

Question 4.3 Construct a triangle of force vectors for the two forces shown in Fig. 4.5 and their equilibrant force.

Bow's notation

To enable a triangle of forces to be constructed easily, use is made of a system of lettering known as Bow's notation.

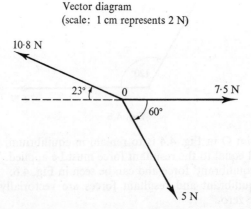

Fig. 4.7 *Reduced from original: do not scale*

26 THE PICTORIAL REPRESENTATION OF FORCES

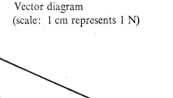
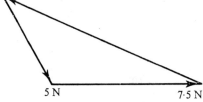

Fig. 4.8 *Reduced from original: do not scale*

Example 4.4 Using Bow's notation, draw a triangle of forces for the 7·5 N, 5 N, and 10·8 N forces shown in Fig. 4.7.

The complete solution to this example is shown in Fig. 4.9.

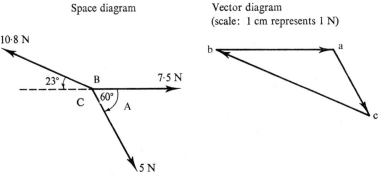

Fig. 4.9 *Reduced from original: do not scale*

The method used in the construction of the vector diagram is as follows:

STEP 1 The spaces between the forces shown in the space diagram (Fig. 4.9) are lettered A, B, and C. The 10·8 N force is called force AC, the 7·5 N force is called force AB, and the 5 N force is called force BC.

STEP 2 Draw the vector *ab* parallel with force AB and to a scale length of 7·5 N. Label *ab* by the side of the vector (Fig. 4.10).

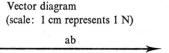

Fig. 4.10 Vector of force AB. *Reduced from original: do not scale*

STEP 3 From the end of vector *ab*, a vector *ac* may then be drawn parallel with force AC. As either the left-hand or the right-hand end of vector *ab* could be end '*a*', and the direction could be either up or down, there are four possible positions for this vector (Fig. 4.11).

THE THEOREM OF THE TRIANGLE OF FORCES   27

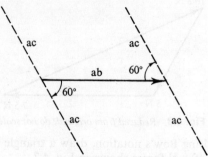

Fig. 4.11 Vector of force AB and possible positions of AC.
Reduced from original: do not scale

STEP 4 Draw from end '*b*,' which could also be either end, a vector *bc* parallel with force BC.

It now becomes clear that two triangles can be formed (Fig. 4.12).

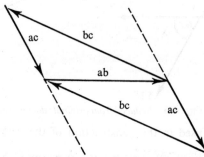

Fig. 4.12 The method by which the three vectors will construct two possible triangles. *Reduced from original: do not scale*

STEP 5 Choose the most convenient triangle and line in boldly. Figures 4.13 and 4.14 show the alternative triangles.

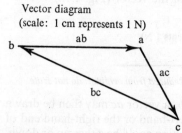

Fig. 4.13 *Reduced from original: do not scale*

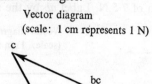

Fig. 4.14 *Reduced from original: do not scale*

28 THE PICTORIAL REPRESENTATION OF FORCES

The corners of the vector triangles can now be lettered:

Point 'a' is common to ac and ab.
Point 'b' is common to ab and bc.
Point 'c' is common to ac and bc.

Practical demonstration of the theorem of the triangle of forces

The theorem of the triangle of forces may be verified with the aid of a forces board, arranged as shown in Fig. 4.15, and the magnitude of forces F_1, F_2, and F_3 noted. The position of the cords are copied onto a backing paper, thus forming a space diagram. From this space diagram, a reasonably closed, triangular-shaped vector diagram may be drawn.

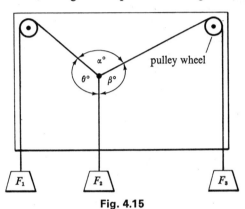

Fig. 4.15

Example 4.5 The angles α, θ, and β, shown in Fig. 4.15, are measured as 100°, 120°, and 140° respectively. If F_2 is 2 kgf, find the value of F_1 and F_3.

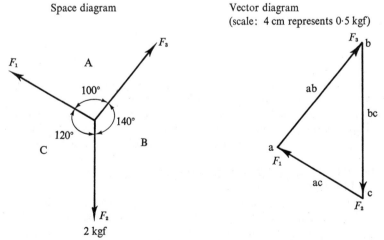

Fig. 4.16 *Reduced from original: do not scale*

From the vector diagram of Fig. 4.16 the forces are:

$$F_1 = 1\cdot 3 \text{ kgf} \quad \text{and} \quad F_3 = 1\cdot 8 \text{ kgf}$$

Question 4.4 If on the forces board shown in Fig. 4.15 the angles $\alpha = 90°$, $\theta = 110°$, and $\beta = 160°$, and if F_2 is 3 kgf, find the value of F_1 and F_3 required to maintain a condition of equilibrium. ($F_1 = 1$ kgf; $F_3 = 2\cdot 8$ kgf)

Concurrency

If a lamina is suspended from the forces board, as shown in Fig. 4.17, and is allowed to settle into a position of equilibrium, then an extension of the lines of the forces as they meet the lamina will intersect at a common point.

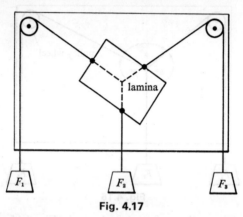

Fig. 4.17

The rule of concurrency may be stated as: *The lines of action of any three forces, acting in the same plane, and causing equilibrium, will intersect at a common point.*

4.5 Centre of mass

The centre of mass, which is commonly called the centre of gravity, is the position within an object at which all the mass may be considered to concentrate.

A practical method of finding the centre of mass of a small flat object is to suspend the object, in turn, from two separate positions on its outer surface. In each case an extension of the line of suspension will pass through the position of the centre of mass. The point at which the lines intersect will determine the position of the centre of mass.

4.6 Practical applications of the theorem of the triangle of forces

Example 4.6 A jib crane is supporting a mass of 1000 kg, as shown in Fig. 4.18. The length of the jib is 11 m and the length of the tie cable is 7 m. The jib and tie fixings at the mast are separated by a distance of 6 m. Find the load in the jib and in the tie cable.

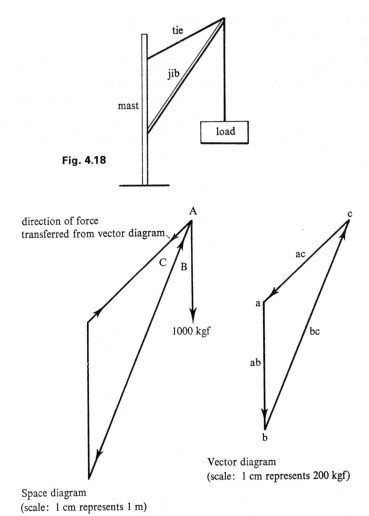

Fig. 4.18

Space diagram
(scale: 1 cm represents 1 m)

Vector diagram
(scale: 1 cm represents 200 kgf)

Fig. 4.19 *Reduced from original: do not scale*

The solution to this problem is shown in Fig. 4.19, in which the force vector *ac* scales 1175 kgf and the force vector *bc* scales 1830 kgf.

The direction of vector *ac* (downwards) is now transferred to the space diagram. In a similar way the direction of vector *bc* (upwards) is also transferred.

As the jib and tie are in equilibrium, the direction of the reactions at the lower end of each must be in reverse. As the reactions within the tie cable are pointing inwards, the external forces on this member must be acting outwards; that is, the tie cable is under a stretching load or tension. Since the reactions of the jib are pointing outwards, its load must be compressive. Thus, the answer to Example 4.6 should be:

PRACTICAL APPLICATIONS OF THE TRIANGLE OF FORCES

The tie cable is under a tensile load of 1175 kgf and the jib is under a compressive load of 1830 kgf.

Question 4.5 A jib crane has the following dimensions: tie cable, 4 m; jib, 7·5 m; distance between fixing points on mast, 5 m. If a mass of 700 kg is supported by the crane, find the loads in the tie and the jib.

(Tie, 570 kgf tension; jib, 1060 kgf compression)

Example 4.7 The pin-jointed framework shown in Fig. 4.20 supports a vertical load of 40 kgf. Find the load in each member of the framework and the upward reactions R_1 and R_2.

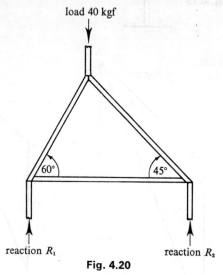

Fig. 4.20

The solution to this problem and the type of loading are shown in Fig. 4.21.

The procedure for the construction of the vector diagram is as follows:

STEP 1 Draw the space diagram and insert all information given. Label the diagram with Bow's notation and identify each joint in the framework.

STEP 2 On examination of joint 1 it will be seen that the magnitude and direction of force AB is given, and as the directions of the forces in members AC and BC must be along the member, a vector triangle, *abc*, may be constructed.

A small sketch is then made of vector triangle *abc* to show the direction of the forces.

STEP 3 The direction of the vectors *ac* and *bc* are now transferred to the space diagram. The forces in members AC and BC at joints 2 and 3 will be opposite in direction, and these can also be shown.

STEP 4 There is now sufficient information to construct the vector triangles for joints 2 and 3—triangles *acd* and *bcd* respectively.

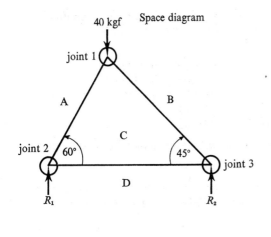
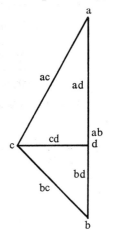
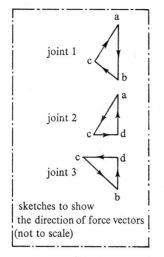
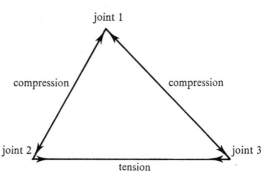

Fig. 4.21 *Reduced from original: do not scale*

By drawing the vector *cd* parallel with member CD, and from point '*c*' on the vector diagram, both triangles are constructed.

Small sketches are made to show the direction of the forces around the vector triangles of joints 2 and 3. It will be noticed that the direction of the forces in the three sketches are not the same, and therefore to attempt to show them on the vector diagram would be most confusing. The answer to Example 4.7 should be stated as follows.

Load in framework members:

> Member AC   29 kgf (compression)
> Member BC   21 kgf (compression)
> Member CD   14·5 kgf (tension)
> Reaction R_1 (AD)   25·5 kgf
> Reaction R_2 (BD)   14·5 kgf

The sum of the reactions ($R_1 + R_2$) should equal the applied load:

$$25·5 + 14·5 = 40 \quad correct$$

Question 4.6 If the vertical load applied to the pin-jointed frame XYZ shown in Fig. 4.22 is 32 kgf, find the load in each member and the reactions, R_1 and R_2.

> [XZ   22·5 kgf (compression)
> XY   22·5 kgf (compression)
> YZ   16 kgf (tension)
> R_1 16 kgf
> R_2 16 kgf]

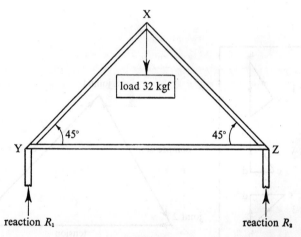

Fig. 4.22

4.7 Polygon of forces

This theorem is similar to the theorem of the triangle of forces, but applies to a number of forces greater than three. The principle of concurrency will, however, no longer be true.

If any number of non-parallel forces, acting in the same plane, cause a condition of equilibrium, then a closed vector figure may be drawn of the forces. The direction of the forces will follow around the vector diagram.

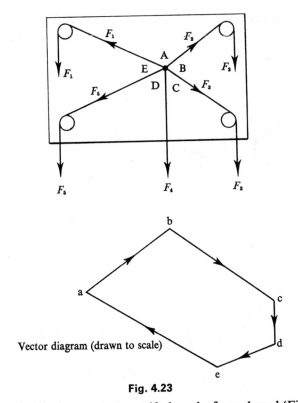

Vector diagram (drawn to scale)

Fig. 4.23

This principle may again be verified on the forces board (Fig. 4.23).

Example 4.8 Find the fifth force that would provide equilibrium to the four forces, acting in the same plane, shown in Fig. 4.24.

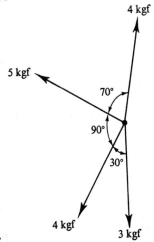

Fig. 4.24

POLYGON OF FORCES 35

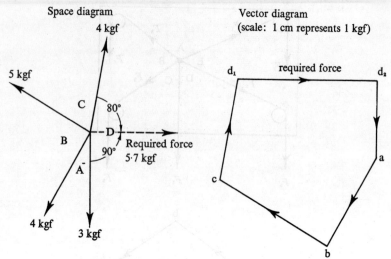

Fig. 4.25 *Reduced from original: do not scale*

SOLUTION From the vector diagram of Fig. 4.25, the required force is 5·7 kgf, acting at 90° to the 3 kilogramme force AD, and at 80° to the 4 kilogramme force CD.

Question 4.7 Find the fourth force required to provide equilibrium to the three forces shown in Fig. 4.26.

(3·3 N; 97° to 3 N force and 63° to 2 N force)

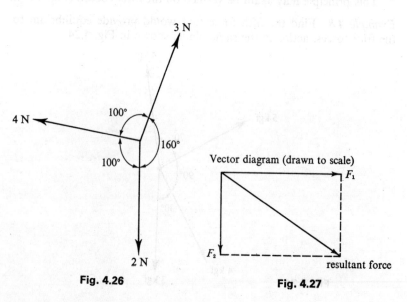

Fig. 4.26

Fig. 4.27

36 THE PICTORIAL REPRESENTATION OF FORCES

4.8 Resolution of forces into right angle components

Figure 4.27 shows two force vectors, at right angles to one another, added together to form a resultant.

If the process is reversed, the resultant vector can be split into the two vectors F_A and F_B, as shown in Fig. 4.28.

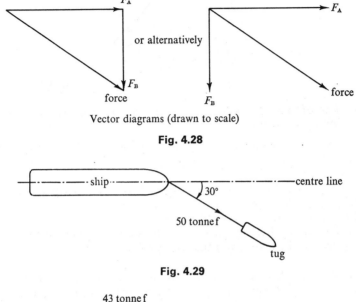

Fig. 4.28

Fig. 4.29

Fig. 4.30 *Reduced from original: do not scale*

Example 4.9 A ship is towed by a tug, the towing cable making an angle of 30° in the horizontal plane with the centre-line of the ship (Fig. 4.29). If there is a 50 tonnef (50 000 kgf) in the tow cable, find the force acting (a) along the centre-line of the ship, and (b) at right angles to the centre-line.

From the vector diagram, Fig. 4.30, the force along the centre-line is 43 tonnef and the force at right angles to the centre-line is 25 tonnef.

Question 4.8 A shelf is supported at one end by a cord which makes an angle of 45° with the shelf, as shown by Fig. 4.31. If the loading of the cord is 3 kgf, find the force (a) at right angles to the shelf and (b) along the line of the shelf. (2·12 kgf; 2·12 kgf)

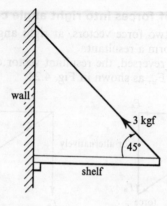

Fig. 4.31

4.9 The effect of the weight of a mass resting on an inclined plane

If a mass is resting on an inclined plane, the weight of the mass, which is its downward force due to the acceleration of gravity, will act as shown in Fig. 4.32.

The downward force may be resolved into two components, one parallel with the inclined plane and the other at right angles to the inclined plane. This is shown in Fig. 4.33.

The force E is holding the mass to the gradient, while the force P is trying to pull the mass down the gradient.

Example 4.10 A mass of 20 kg rests on a gradient, as shown in Fig. 4.34. Find the force attempting to slide the mass down the gradient.

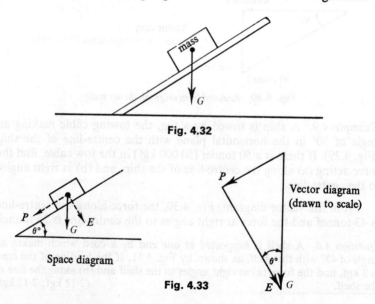

Fig. 4.32

Space diagram

Vector diagram (drawn to scale)

Fig. 4.33

38 THE PICTORIAL REPRESENTATION OF FORCES

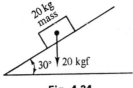

Fig. 4.34

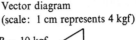

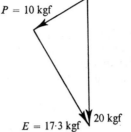

Fig. 4.35 *Reduced from original: do not scale*

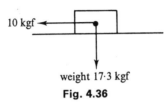

Fig. 4.36

Resolving the forces:
From the vector diagram, Fig. 4.35, the force down and parallel to the gradient is 10 kgf.

The condition, shown by Fig. 4.34, could be reproduced by a mass of 17·3 kg resting on a horizontal surface, plus a horizontal force of 10 kgf acting on the mass (see Fig. 4.36).

Let a friction force of 10 kgf be acting on the mass in opposition to the applied force of 10 kgf.

The coefficient of friction for the surface is

$$\mu = \frac{\text{Friction force}}{\text{Downward force}} = \frac{10}{17\cdot 3} = 0\cdot 577$$

If the tangent of the gradient angle is found from tables, then

$$\tan 30° = 0\cdot 577$$

Critical angle of gradient

When the tangent of the angle of gradient is equal to the coefficient of friction, then this angle is critical. If the gradient becomes any steeper the mass will begin to slide.

Example 4.11 The coefficient of friction between a surface and a mass of 50 kg is 0·3. Find (a) the friction force and (b) the maximum angle of gradient that the mass could rest upon without sliding, and (c) verify the results by considering an equivalent mass resting on a similar horizontal surface.

THE EFFECT OF THE WEIGHT OF A MASS

(a) When the mass is resting on a horizontal surface the downward force will be 50 kgf.

$$\text{Friction force} = \mu \times \text{Downward force}$$
$$= 0{\cdot}3 \times 50 = 15 \text{ kgf}$$

(b) The critical angle of gradient will be the angle whose tangent is 0·3, that is 16·7°.

The answers to (a) and (b) could also be approximated by constructing the diagrams shown in Figs. 4.37 and 4.38.

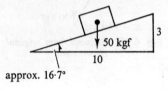

Fig. 4.37

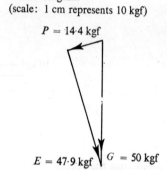

Fig. 4.38 *Reduced from original: do not scale*

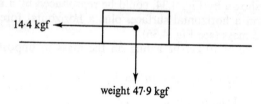

Fig. 4.39

(c) If the mass is now considered to be an equivalent mass of 47·9 kg resting on an equivalent horizontal surface, and with an applied horizontal force of 14·4 kgf (Fig. 4.39), then

$$\frac{14{\cdot}4}{47{\cdot}9} = 0{\cdot}3 \quad (verified)$$

Question 4.9 A hill has a slope of 1 in 20 (a vertical rise of 1 metre in a distance along the slope of 20 metres). A car is found to roll down the hill at a constant velocity. Find the coefficient of friction between the car and the road.   (0·05)

Additional questions

4.10 (a) Construct vector diagrams and find the resultant and equilibrant of the forces shown in Fig. 4.40.

(b) Check your answers by drawing in each case a triangle of forces.

4.11 (a) Draw the vectors of the forces shown in Fig. 4.41 and find their equilibrant. (*Hint:* Add, vectorially, two of the forces then add their resultant to the third force.)
(b) Check your solution by constructing a polygon of forces in each case.

4.12 A lamp fitting of mass 10 kg, is suspended midway along a cord 3 m in length, one end of which is attached to the ceiling and the other to the wall, as shown in Fig. 4.42. Find the tension in each section of the cord.

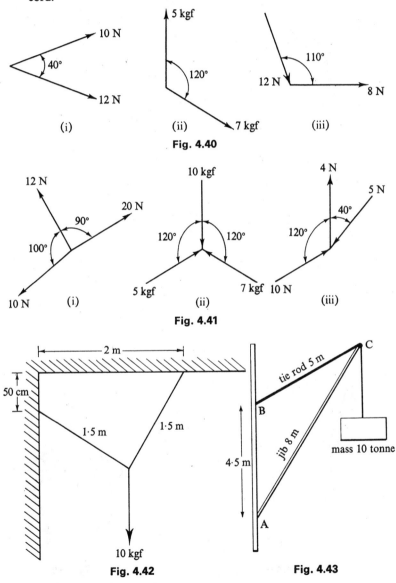

Fig. 4.40

Fig. 4.41

Fig. 4.42

Fig. 4.43

ADDITIONAL QUESTIONS

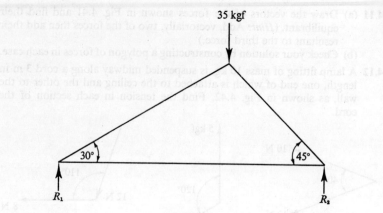

Fig. 4.44

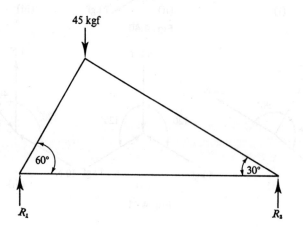

Fig. 4.45

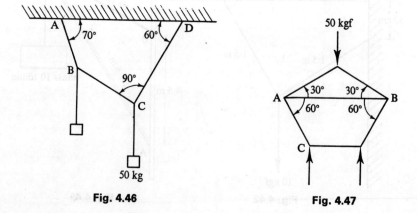

Fig. 4.46   **Fig. 4.47**

42 THE PICTORIAL REPRESENTATION OF FORCES

4.13 A mass of 10 tonne (Mg) is supported as shown in Fig. 4.43 by a jib crane, in which the vertical member AB is 4·5 m long, the jib AC is 8 m long, and the tie rod BC is 5 m long. Determine the tension in the tie rod and the thrust in the jib.

4.14 For the roof truss shown in Fig. 4.44 find the following information: (a) the reactions R_1 and R_2; (b) the load in each member; and (c) the type of loading of each member.

4.15 For the framework shown in Fig. 4.45 find (a) the loading of each member and (b) the reactions at the framework supports.

4.16 Masses are suspended from two points B and C on a rope ABCD, as shown in Fig. 4.46. The ends A and D are on the same horizontal level. Find the tension in each section of the rope, and the mass suspended from the point B. (Based upon C.G.L.I.)

4.17 A load of 50 kgf is supported by a pin-jointed frame as shown in Fig. 4.47. Find graphically the loads in the members AB and AC.
(Based upon C.G.L.I.)

4.18 A load of 100 kgf is supported by two wires attached to eye bolts, one in the wall and one in the roof, as shown in Fig. 4.48. Find the tension in each wire. (Based upon C.G.L.I.)

4.19 Figure 4.49 shows four coplanar forces acting at one point. Find (a) the horizontal and vertical components of the 2 N force; (b) the resultant of the four forces; and (c) the acceleration produced if these are the only forces acting on a mass of 2 kg. (C.G.L.I.)

4.20 A force of 20 kgf is used to tow an empty trailer of mass 200 kg up an incline of one in twenty (i.e. 1 unit rise in 20 units of slope distance). Find (a) the components of the weight of the trailer (i) parallel to, and (ii) perpendicular to the ground; (b) the acceleration of the empty trailer; and (c) the maximum mass that can be placed in the trailer and still allow the towing force to move it from rest. (C.G.L.I.)

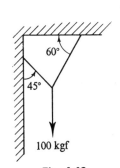

Fig. 4.48

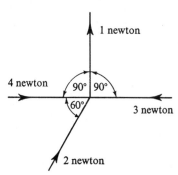

Fig. 4.49

5 Moments and Parallel Forces

Consider the freely pivoted wheel and axle shown in Fig. 5.1.

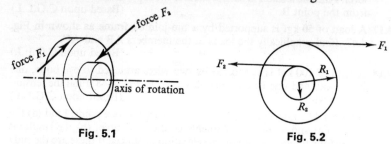

Fig. 5.1   Fig. 5.2

Let the wheel radius be designated R_1 and the axle radius R_2.

If the wheel were held by one hand providing force F_1, and the axle held by the other hand providing force F_2 in opposition to F_1, then to keep the wheel and axle in equilibrium force F_2 would need to be much larger than force F_1.

Alternatively, these forces could be provided by cords wound around the circumference of the wheel and axle (Fig. 5.2).

By inserting spring balances in the cords, the values of F_1 and F_2 could be found. For the wheel and axle to remain in equilibrium, the ratio of F_1 to F_2 would need to be the same as the ratio of R_2 to R_1, that is

$$F_1 : F_2 :: R_2 : R_1$$

F_1 is said to vary with F_2 as R_2 varies with R_1. From this ratio

$$\frac{F_1}{F_2} = \frac{R_2}{R_1}$$

and

$$F_1 \times R_1 = F_2 \times R_2$$

As these cords are wound on the circumference of the wheel and axle, forces F_1 and F_2 must be applied as tangents to the wheel and axle; that is, at right angles to the radius.

The products $F_1 \times R_1$ and $F_2 \times R_2$ are called **moments**. $F_1 \times R_1$ is the moment of F_1 about the pivot, and $F_2 \times R_2$ is the moment of F_2 about the pivot.

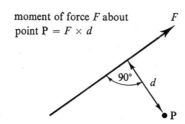

moment of force F about
point P = F × d

Fig. 5.3

5.1 Definition of a moment

The moment of a force about a point is the product of that force and the perpendicular distance between the force and the point.

This definition may be clarified by a simple sketch, Fig. 5.3, which is an excellent way of remembering the essential details of the definition.

Moment units

In the SI system the absolute unit used is the newton-metre (N m). Other possible units would be kilogramme-force metre (kgf m), newton-centimetre (N cm), etc. It should be noted that the unit of force is always placed first.

Example 5.1 A desk lid is lifted by applying a force of 1 kgf at the front edge of the lid. If the distance between this edge and the hinge is 0·5 m, find the moment of the force about the hinge in kilogramme-force metres and newton-metres, (a) if the force is applied, as shown in Fig. 5.4, at right angles to the lid and (b) if the force is applied, as shown in Fig. 5.5, at an angle of 45° to the lid.

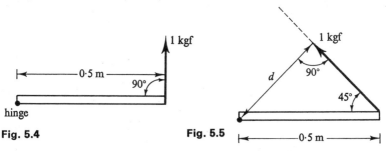

Fig. 5.4 **Fig. 5.5**

(a) Moment of 1 kgf about the hinge is

Force × Perpendicular distance = 1 × 0·5 = 0·5 kgf m
or   = 1 × 9·81 × 0·5 = 4·905 N m

(b) The distance d can be found either by drawing a space diagram to scale, or by trigonometry:

$$\frac{d(\text{opp.})}{0\cdot5(\text{hyp.})} = \sin 45° = 0\cdot7071$$

DEFINITION OF A MOMENT **45**

$$d = 0\cdot5 \times 0\cdot7071 = 0\cdot3536 \text{ m}$$

The moment of 1 kgf about the hinge is

Force × Perpendicular distance = $1 \times 0\cdot3536 = 0\cdot3536$ kgf m

or $\quad\quad\quad\quad = 1 \times 9\cdot81 \times 0\cdot3536 = 3\cdot46$ N m

Question 5.1 (a) A shelf is supported by a vertical cord as shown in Fig. 5.6. If the load in the cord is 3 kgf, find the moment in newton-metres of the force about the point P.

(b) The angle of the cord to the shelf is now altered to 30°. If the moment about point P remains unchanged, find the load in newtons that must be applied to the cord. (8·829 N m; 58·86 N)

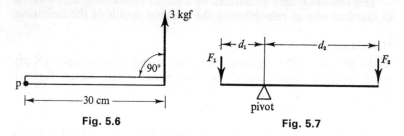

Fig. 5.6   Fig. 5.7

5.2 Bodies in equilibrium

If a body is in a state of equilibrium the algebraic sum of the moments about any point must equal zero.

Figure 5.7 shows a balance which is in a state of equilibrium.

The moment of F_1 about the pivot is $F_1 \times d_1$ and is called an anti-clockwise moment as the direction of the force F_1 is tending to turn the balance in an anti-clockwise direction. In a similar way the moment of F_2 about the pivot is $F_2 \times d_2$, clockwise.

Alternatively, the condition for a body to be in a state of equilibrium can be expressed as 'the sum of the clockwise moments about any point must equal the sum of the anti-clockwise moments', that is,

$$F_1 \times d_1 = F_2 \times d_2$$

5.3 Reaction at the pivot

In Fig. 5.7 the pivot is supporting the parallel forces F_1 and F_2. The reaction at the pivot must therefore be equal to $F_1 + F_2$ but must be opposite in direction; that is, upwards.

In general, for equilibrium the algebraic sum of the applied load must be equal to the algebraic sum of the reactions.

If more than two forces are acting on the body, as shown in Fig. 5.8, the algebraic sum of the forces will be

$$F_1 + F_3 - F_2 - F_4 = \text{Reaction at the pivot}$$

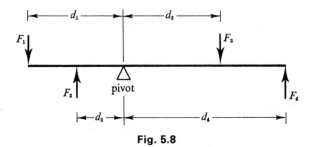

Fig. 5.8

If $F_1 + F_3$ is greater than $F_2 + F_4$ the reaction will be upwards, and if $F_2 + F_4$ is greater than $F_1 + F_3$ the reaction will be downwards.

Taking moments about the pivot of the beam shown in Fig. 5.8:
The sum of the anti-clockwise moments is

$$(F_1 \times d_1) + (F_4 \times d_4)$$

The sum of the clockwise moments is

$$(F_2 \times d_2) + (F_3 \times d_3)$$

The reaction force at the pivot has no moment here as it is acting at zero distance from the pivot and will be $R \times 0 = 0$.

For equilibrium the sum of the anti-clockwise moments must equal the sum of the clockwise moments:

$$(F_1 \times d_1) + (F_4 \times d_4) = (F_2 \times d_2) + (F_3 \times d_3)$$

Example 5.2 In the balance shown in Fig. 5.9, W_2 is 100 gf, d_2 is 30 cm, and d_1 is 10 cm. Find, in grammes-force and newtons (a) the value of W_1 to keep the balance in equilibrium and (b) the reaction R at the pivot.

(a) Taking moments about the pivot:

$W_1 \times d_1 = W_1 \times 10 = 10W_1$ gf cm (anti-clockwise moment)
$W_2 \times d_2 = 100 \times 30 = 3000$ gf cm (clockwise moment)
(Moment of R about pivot $= R \times 0 = 0$)

For equilibrium the sum of the anti-clockwise moments must equal the sum of the clockwise moments:

$$10W_1 = 3000$$
$$W_1 = 300 \text{ gf}$$

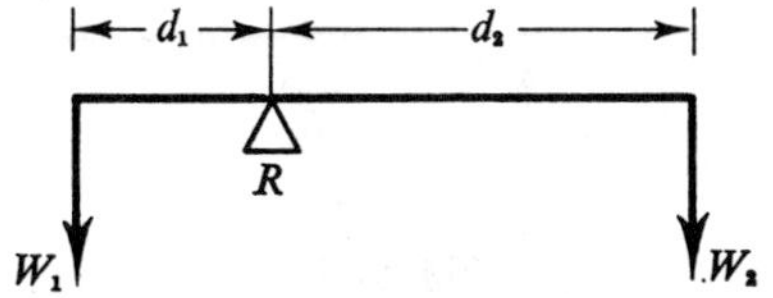

Fig. 5.9

or
$$W_1 = \frac{300}{1000} \text{ kgf}$$
$$= \frac{300}{1000} \times 9\cdot81 \text{ N}$$
$$= 2\cdot943 \text{ N}$$

(b) The sum of the downward forces is
$$W_1 + W_2 = 100 + 300 = 400 \text{ gf}$$

The reaction at R must therefore be

$$400 \text{ gf upwards}$$

or, in newtons,
$$R = \frac{400}{1000} \times 9\cdot81 = 3\cdot924 \text{ N}$$

Question 5.2 If, in Fig. 5.9, W_1 is 0·5 kgf, d_1 is 10 cm, and d_2 is 20 cm, find in kilogrammes-force and newtons (a) the value of W_2 required to maintain equilibrium and (b) the reaction at the pivot.

(0·25 kgf, 2·45 N; 0·75 kgf, 7·35 N)

Question 5.3 If, in Fig. 5.9, W_1 is 50 gf, d_1 is 10 cm, and W_2 is 40 gf, find (a) the distance from the pivot that W_2 must be placed to maintain equilibrium and (b) the reaction at the pivot. (12·5 cm; 90 gf)

Example 5.3 In the balance shown in Fig. 5.10, lengths $d_1 = 4$ cm, $d_2 = 1$ cm, and $d_3 = 3$ cm, and loads $W_1 = 5$ N and $W_2 = 2$ N. Find (a) the value of W_3 necessary for equilibrium and (b) the reaction at the pivot.

(a) Taking moments about the pivot:

The sum of the anti-clockwise moments is
$$W_1 \times d_1 = 5 \times 4 = 20 \text{ N cm}$$

The sum of the clockwise moments is
$$(W_2 \times d_2) + (W_3 \times d_3) = 2 \times 1 + W_3 \times 3$$
$$= 2 + 3W_3 \text{ N cm}$$

For equilibrium the sum of the anti-clockwise moments must equal the sum of the clockwise moments:

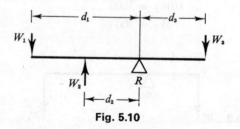

Fig. 5.10

$$2 + 3W_3 = 20$$
$$3W_3 = 18$$
$$W_3 = 6 \text{ N}$$

(b) Downward force = $W_1 + W_3 = 5 + 6 = 11$ N
Upward force = $W_2 = 2$ N
Resultant force = $11 - 2 = 9$ N downwards

Therefore the reaction must be opposite; that is, 9 N upwards.

Question 5.4 In Fig. 5.10, W_1 is 10 gf, W_3 is 4 gf, and the reaction at the pivot is 6 gf upwards. If d_1 is 12 cm and d_3 is 20 cm, find (a) the value of W_2 and (b) its position if equilibrium is to be maintained.

(8 gf; 5 cm from pivot)

Example 5.4 In Fig. 5.11 the load W_1 is 10 N, W_2 is 20 N, W_3 is 20 N,

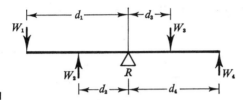

Fig. 5.11

W_4 is 15 N, and the distance d_1 is 5 cm, d_2 is 3 cm, and d_4 is 10 cm. Find (a) the position of W_3 and (b) the magnitude and direction of the reaction R at the fulcrum if equilibrium is to be maintained.

(a) Taking moments about the fulcrum:
The sum of the anti-clockwise moments is
$$(W_2 \times d_2) + (W_3 \times d_3) = (20 \times 3) + (20 \times d_3)$$
$$= 60 + 20d_3 \text{ N cm}$$

The sum of the clockwise moments is
$$(W_1 \times d_1) + (W_4 \times d_4) = (10 \times 5) + (15 \times 10)$$
$$= 50 + 150 = 200 \text{ N cm}$$

For equilibrium the sum of the anti-clockwise moments must equal the sum of the clockwise moments:
$$60 + 20d_3 = 200$$
$$20d_3 = 140$$
$$d_3 = 7 \text{ cm}$$

(b) For equilibrium the total downward force must equal the total upward force:

Downward load = $W_1 + W_3 = 10 + 20 = 30$ N
Upward load = $W_2 + W_4 = 20 + 15 = 35$ N
Resultant load = $35 - 30 = 5$ N upwards

The reaction at the fulcrum must be 5 N downwards and not as diagrammatically shown in Fig. 5.11.

Question 5.5 In Fig. 5.11 the load W_1 is 40 newton, W_2 is 30 newton, W_4 is 10 newton, and the distance d_1 is 5 cm, d_2 is 2 cm, d_3 is 4 cm, and d_4 is 10 cm. Find (a) the value of W_3 and (b) the magnitude and direction of the reaction at the pivot for equilibrium to be maintained.

(60 N; 60 N upwards)

The moment principle is widely used in engineering and many of the machines considered in Chapter 7, as well as common devices such as pliers, spanners, wrenches, and even nutcrackers, operate by this principle.

Question 5.6 Name three other devices which use the principle of moments.

In practice moments also have disadvantages. For example, the hinge of a door can easily be distorted if something is trapped in the hinged end of the door.

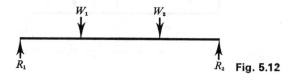
Fig. 5.12

If a refrigerator door is 50 cm wide and a screwdriver 5 mm thick is trapped in the hinged side of the door, a force of 2 kgf applied at the handle of the door will exert a force on the hinge of 200 kgf, probably with disastrous results. This can be calculated as follows:

$$\text{Moment of force on hinge} = \text{Moment of applied force}$$

$$\text{Force} \times 0.5 = 2 \times 50$$

Thus,
$$\text{Force on hinge} = \frac{50 \times 2}{0.5} = 200 \text{ kgf}$$

5.4 Beam supports

In practice a beam must have at least two supports, a typical arrangement of forces being shown in Fig. 5.12. W_1 and W_2 are called point loads; that is, they are acting at one point rather than over an area, which would be the normal condition.

Any type of load may be indicated by a point force if the position of the force is taken through the centre of mass of the load.

5.5 Centre of mass

The **centre of mass** is the position at which all the mass of a body can be considered to be concentrated for calculation purposes. The weight of a distributed load may therefore be turned into a point load if the centre of mass of the load is known or can be calculated.

An example of a distributed load is the weight of the beam itself. If the beam is of uniform cross-section and material, then the centre of mass will be at the mid-point of the beam.

5.6 Calculation of the reactions at the beam supports

The method used to calculate the reactions at the beam supports is to take moments about the application point of each support in turn.

Example 5.5 Find the reactions R_1 and R_2 to provide a condition of equilibrium for the beam shown in Fig. 5.13. The weight of the beam may be ignored.

Taking moments about R_1:
The sum of the anti-clockwise moments is

$$R_2 \times (2 + 3 + 1) = 6R_2 \text{ kgf m}$$

The sum of the clockwise moments is

$$5 \times 2 + 6 \times (2 + 3) = 10 + 30 = 40 \text{ kgf m}$$

For equilibrium the sum of the anti-clockwise moments must equal the sum of the clockwise moments:

$$6R_2 = 40$$

therefore, $\qquad R_2 = 6\cdot66 \text{ kgf}$

Taking moments about R_2:
The sum of the anti-clockwise moments is

$$6 \times 1 + 5 \times (3 + 1) = 6 + 20 = 26 \text{ kgf m}$$

The sum of the clockwise moments is

$$R_1 \times (1 + 3 + 2) = 6R_1 \text{ kgf m}$$

Again, for equilibrium, $6R_1 = 26$

therefore, $\qquad R_1 = 4\cdot33 \text{ kgf}$

As a check the downward forces must equal the upward forces:

$$R_1 + R_2 = 5 + 6$$
$$4\cdot33 + 6\cdot66 = 5 + 6$$

Thus,
$$R_1 = 4\cdot33 \text{ kgf} \quad \text{and} \quad R_2 = 6\cdot66 \text{ kgf}$$

In order to check that the beam can withstand the shearing action caused by its load, the shear force may be calculated:

Between reaction R_1 and the 5 kgf load the shear force will be equal to the value of reaction R_1, that is, 4·33 kgf.

At the 5 kgf point load the shear force will be zero, as at that point the upward and downward loads are equal. Between the 5 kgf and 6 kgf loads the shear force will be $5 - 4\cdot33 = 0\cdot66$ kgf. At the 6 kgf point load the shear force will again be zero.

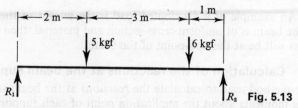

Fig. 5.13

Between the 6 kgf load and reaction R_2 the shear force will be $6 + 0.66 = 6.66$ kgf, which will be seen to be equal to reaction R_2.

Question 5.7 If the beam shown in Fig. 5.13 has a mass of 3 kg and is of uniform cross-section find the reactions R_1 and R_2. ($5\frac{5}{8}$ kgf; $8\frac{1}{8}$ kgf)

Example 5.6 (a) Find the reactions R_1 and R_2 for the beam shown in Fig. 5.14. The beam has a mass of 4 kg.

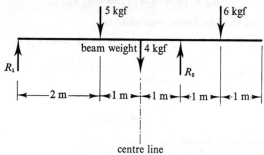

Fig. 5.14

(b) Find the shear force at a position 1 m from reaction R_1.

Taking moments about R_1:

The sum of the anti-clockwise moments is

$$R_2 \times (2 + 1 + 1) = 4R_2 \text{ kgf m}$$

The sum of the clockwise moments is

$$5 \times 2 + 4 \times (2 + 1) + 6 \times (2 + 1 + 1 + 1) = 10 + 12 + 30 = 52 \text{ kgf m}$$

For equilibrium:
$$4R_2 = 52$$
therefore,
$$R_2 = 13 \text{ kgf}$$

Taking moments about R_2:

The sum of the anti-clockwise moments is

$$4 \times 1 + 5 \times (1 + 1) = 4 + 10 = 14 \text{ kgf m}$$

The sum of the clockwise moments is

$$6 \times 1 + R_1 \times (1 + 1 + 2) = 6 + 4R_1$$

For equilibrium:
$$6 + 4R_1 = 14$$
therefore,
$$4R_1 = 8$$
and
$$R_1 = 2 \text{ kgf}$$

52 MOMENTS AND PARALLEL FORCES

As a check, the downward forces must equal the upward forces for equilibrium.

The sum of the downward forces is

$$5 + 4 + 6 = 15 \text{ kgf}$$

The sum of the upward forces is

$$R_1 + R_2 = 2 + 13 = 15 \text{ kgf}$$

thus,

$$R_1 = 2 \text{ kgf} \quad \text{and} \quad R_2 = 13 \text{ kgf}$$

The shear force loading on a beam must be increased by the beam's own weight.

The shear force 1 m from reaction R_1 will be reaction R_1, which is 2 kgf upwards, less the weight of the beam between reaction R_1 and the point under consideration, that is $1 \times \frac{4}{6}$ kgf downwards.

Therefore, the shear force 1 m from reaction R_1 is

$$2 - \tfrac{2}{3} = 1\tfrac{1}{3} \text{ kgf}$$

Question 5.8 A 10 m long uniform beam has a mass of 5 kg. The beam is supported at the right-hand end and 2 m from the left-hand end. A point load of 6 kgf is placed at the extreme left-hand end. Find the reactions at the supports, and the shear force 1 m from the left-hand end of the beam.

(L/h 10·625 kgf; R/h 0·375 kgf; 6·5 kgf)

5.7 Parallel forces

The theorems of the triangle and polygon of forces (Chapter 4) excluded forces which are parallel.

Consider the two parallel forces A and B acting in the same plane and in the same direction (Fig. 5.15). Let these forces act on an imaginary beam of zero mass (Fig. 5.16). To balance these forces—that is, to

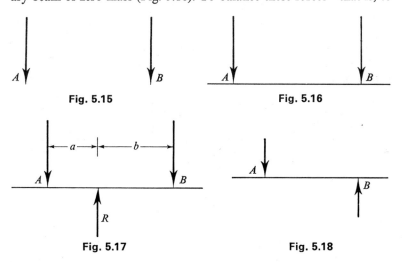

Fig. 5.15   **Fig. 5.16**

Fig. 5.17   **Fig. 5.18**

keep the beam in equilibrium—a single reaction R would need to be positioned between A and B. The value of this reaction would be $A + B$ and its direction would be upwards (Fig. 5.17).

The position of this reaction may be found by taking moments about A and B in turn.

Taking moments about A: for equilibrium

$$Ra = B(a + b)$$

therefore,
$$R = \frac{B(a + b)}{a} \quad (5.1)$$

Taking moments about B: for equilibrium

$$Rb = A(a + b)$$

therefore,
$$R = \frac{A(a + b)}{b} \quad (5.2)$$

From Equations (5.1) and (5.2)

$$R = \frac{B(a + b)}{a} = \frac{A(a + b)}{b}$$

thus,
$$\frac{B(a + b)}{a} = \frac{A(a + b)}{b}$$

$$\frac{B}{a} = \frac{A}{b}$$

and
$$\frac{A}{B} = \frac{b}{a}$$

That is, the ratio of the distances between the forces and the reaction is the inverse of the magnitude of the two forces.

Let force B be reversed and let A be the greater of the two forces (Fig. 5.18).

The reaction R will be $A - B$ and must be upwards. The position of this reaction must be as shown in Fig. 5.19 if equilibrium is to be maintained.

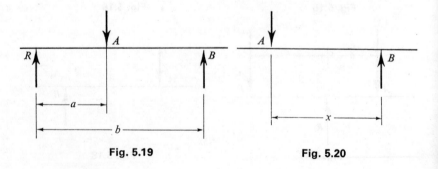

Fig. 5.19   Fig. 5.20

Taking moments about A: for equilibrium

$$R \times a = B \times (b - a)$$

therefore, $$R = \frac{B(b - a)}{a} \qquad (5.3)$$

Taking moments about B: for equilibrium

$$R \times b = A \times (b - a)$$

therefore, $$R = \frac{A(b - a)}{b} \qquad (5.4)$$

From Equations (5.3) and (5.4)

$$R = \frac{B(b - a)}{a} = \frac{A(b - a)}{b}$$

thus, $$\frac{B(b - a)}{a} = \frac{A(b - a)}{b}$$

$$\frac{B}{a} = \frac{A}{b}$$

and, $$\frac{A}{B} = \frac{b}{a}$$

and the result is as before.

The couple

The special case of equal but opposite parallel forces is called a **couple**. There cannot be a reaction to a couple as $R = A - B = 0$. A couple, however, will have a moment (Fig. 5.20).

Taking moments about A: anti-clockwise moment is

$$A \times 0 + B \times x = Bx$$

Taking moments about B: anti-clockwise moment is

$$B \times 0 + A \times x = Ax$$

But as $A = B$,

$$Bx = Ax$$

Therefore the moment of a couple is the product of one of the forces and the distance between the forces.

5.8 Torque

If a force F_1 is applied to the wheel of Fig. 5.2 at a radius of R_1, then the force will try to rotate the wheel against the opposition of F_2.

As was previously described, both the force and the radius are equally important as it is their product that is effective. The combination of force and radius as applied to a wheel is thus given the special name, **torque**.

Torque is the turning moment of a wheel.

Torque = Force (applied as a tangent) × Radius

The units are those of the moment; that is, the newton-metre, N m.

A wheel will not necessarily revolve when a torque is applied to it as their may also be an opposing torque; for example, the opposing torque of friction.

Torque will again be referred to in the chapters on machines and rotary movement.

Additional questions

5.9 A tyre lever is inserted into the rim of a wheel and a force of 80 N is applied at the end of the lever arm. If the overall length of the lever is 25 cm and the pivot is 2 cm from the end, find the force exerted on the tyre.

5.10 A 6 m long balance, weighing 300 N, is pivoted 2 m from its left-hand end. A point force is applied at the extreme left-hand end to keep the balance in a horizontal position. Find (a) the value of the force required and (b) the reaction at the pivot.

5.11 A 3 m long beam of mass 10 kg is hinged at one end. At the other end it is supported by a wire which makes an angle of 45° with the beam. Find the tension in the wire.

5.12 A lever is used to terminate two wires as shown in Fig. 5.21. If the tension in wire A is 500 gf, find in (a) grammes-force and (b) newtons, the tension required in wire B to maintain equilibrium.

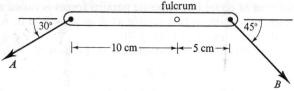

Fig. 5.21

5.13 An aluminium tube 5 cm long of negligible mass is supported at its ends by two reactions. A point load of 10 N is placed 2 cm from one support and 3 cm from the other support. Find the reaction at each support.

5.14 The tube in Question 5.13 is replaced by a tube having a weight of 5 N. Find the reaction at each support.

5.15 A girder 10 m long has a mass of 1 kg per metre of its length. It is supported at one end and at 3 m from the other end. Find the reaction at the supports.

5.16 What is the greatest load that can be placed at the extreme unsupported end of the girder in Question 5.15 before the beam will overbalance? Under these conditions find the reactions at the supports.

5.17 A constantly horizontal force of 5 N is applied to the edge of a cube of side 6 cm and the cube is found to commence to turn over. Find the moment of the applied force about the point of pivot (a) as turning commences and (b) when the cube has turned through an angle of 45°.

5.18 Figure 5.22 shows a 2 m long light fitting having an equally distributed load of 3 kgf. It is suspended by two chains 1 m apart. The starter equipment is placed at the extreme left-hand end and consists of a 30 cm long equally distributed mass of 0·6 kg. Find the tension in each of the chains.

5.19 Figure 5.23 shows a horizontal bar, CD, supported by a vertical wire, AB, to which it is rigidly attached. A couple of 10 N m is needed to produce a 360° twist in the wire. What angle of twist is produced by the system of horizontal forces shown? State briefly what would happen if the 10 N force were removed. (C.G.L.I.)

5.20 State the principle of moments.

The uniform bar shown in Fig. 5.24 weighs 10 kgf and is used to raise a load of 100 kgf. Find (a) the force F required to lift the load, (b) the force exerted by the bar on the fulcrum, and (c) the shear force at the centre of the bar. (C.G.L.I.)

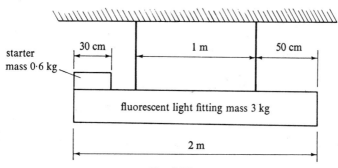

Fig. 5.22

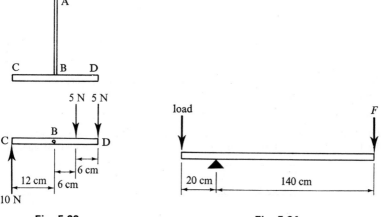

Fig. 5.23　　　　　　　　　　Fig. 5.24

6 Work, Energy, and Power

6.1 Meaning of the term 'work'

When a saw is pushed across a log of wood a relatively large force must be applied, but when it is pulled back the effort required is small. The majority of the **work** of cutting the wood is done on the forward stroke, and it is during this stroke that most energy is expended.

The amount of work done depends upon the amount of force applied and the distance of movement.

Work done is the product of the applied force and the distance of movement in the direction of the force (Fig. 6.1).

The symbol used to denote work is W, the unit of work (energy) in the SI system is the **joule**.

The unit of energy called the joule is the work done when the point of application of a force of one newton is displaced through a distance of one metre in the direction of the force (B.S. 3763).

$$\text{Work} = \text{Force} \times \text{Distance}$$
$$W = F \times l$$

where W is expressed in joules (J), F in newtons (N), and l in metres (m).

Example 6.1 A body is moved a distance of 10 cm by a force of 20 N acting in the direction of movement. Find the work done.

$$W = F \times l$$

The force is 20 N and the distance is $\frac{10}{100}$ m, or 0·1 m; therefore

$$W = 20 \times 0·1 = 2 \text{ J}$$

Question 6.1 A force of 100 N moves a mass through a distance of 5 m in the direction of the force. Find the work done. (500 J)

6.2 Energy

Before work can be done, **energy** must be available. Energy and work are therefore measured in the same units.

Forms of energy

The ultimate source of energy is the sun, and thus energy is received in the form of 'heat'. This energy may be converted into **potential energy**,

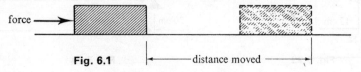

Fig. 6.1

which is the energy of position, and into **kinetic energy,** which is the energy of movement and from which **electrical energy** may be generated.

Theorem of the conservation of energy

Energy is never lost but is converted from one form to another. The **theorem of the conservation of energy** may be explained by reference to Fig. 6.2 which shows the principle of a hydroelectric scheme.

The sun, by the use of heat energy, evaporates water from the sea and causes clouds to form. As these clouds are lifted over mountainous land, heavy rainfall will occur causing lakes to form either naturally or assisted by the construction of dams. This water is higher than sea level and thus will possess energy of height, or potential energy.

The water is released as required and the potential energy changes into kinetic energy—that is, the energy of movement. The water passes through turbines at an electrical generating power station, causing alternators to be rotated and thus generating 'electrical' energy. The electricity is then distributed to the electricity consumer who may use the energy to drive lathes, heat workshops, light offices, and so on, thus all the energy eventually turns back into heat.

At each stage of the conversion process energy is said to be 'lost' due to inefficiency, meaning that this energy is out of human control.

Heat energy

If a mass of 1 kg of water is heated so that its temperature rise is 1 degree Kelvin, then 4186·8 joules of heat energy have been used.

The figure 4186·8 is often taken as 4187 and can even be approximated to 4200.

The degree Kelvin (K) is the temperature scale used in the SI system of units. The freezing point of water is 273 K and boiling point of water is 373 K. As the freezing point of water on the Celsius, or Centigrade, temperature scale is 0°C and the boiling point is 100°C, a rise of 1 K is equal to a rise of 1°C.

In the centimetre-gramme-second (CGS) unit system the heat energy unit is the **calorie,** which is a mass of 1 g of water heated 1°C. To convert this unit, joules-equivalent is used; that is, '1 calorie = 4·1868 J'. As 1 kilocalorie (kcal) = 1000 calories, 1 kcal = 4186·8 J, and this corresponds to the figure used in the SI value of heat energy.

Example 6.2 A mass of 460 g of water is heated from 20°C to 70°C. Find the heat energy required.

$$\text{Heat energy} = \text{Mass of water} \times \text{Rise in temperature}$$
$$= \frac{460}{1000} \times (70 - 20) = 0.46 \times 50 = 23 \text{ kcal}$$
$$= 23 \times 4186.8 = 96\,300 \text{ J}$$

Question 6.2 A mass of 2 kg of water is cooled from 90°C to 30°C. Find the loss of heat from the water. (502 400 J)

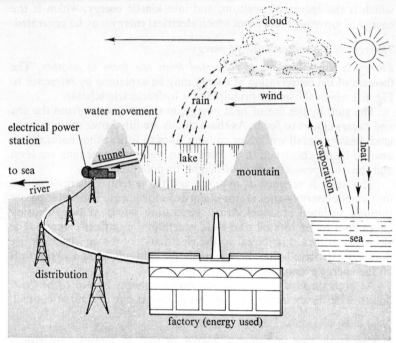

Fig. 6.2

Potential energy

If a mass of M kilogrammes is lifted a vertical distance of h metres at a constant velocity against the acceleration of gravity, then from Newton's second law of motion the force required will be $F = M \times g$.

The work done in the lifting operation is

$$W = F \times h = m \times g \times h$$

Thus Mgh is the potential energy gained by the mass.

Example 6.3 A mass of 10 kg is lifted a height of 4 m. Find the potential energy gained by the mass.

$$W = mgh = 10 \times 9{\cdot}81 \times 4 = 392{\cdot}4 \text{ J}$$

Question 6.3 A mass of 50 kg is dropped through a vertical distance of 10 m. Find the work done on impact with the ground. Into what form would the potential energy be transformed? (4905 J; kinetic then heat)

Kinetic energy

When a mass is moving, the energy of its movement is proportional to its mass and to the square of its velocity.

$$\text{Kinetic energy} = \tfrac{1}{2} \times \text{Mass} \times (\text{Velocity})^2$$

$$\begin{array}{ccc} \text{J} & \text{kg} & \text{m/s} \\ W & = \tfrac{1}{2} \times m & \times \quad v^2 \end{array}$$

60 WORK, ENERGY, AND POWER

This formula may be simply proved:

$$\text{Work, } W = F \times l$$

and since Force $(F) = m \times a$,

$$\text{Work, } W = m \times a \times l$$

With reference to Fig. 1.4 and Example 1.3 the acceleration a is given by the gradient of the graph $a = v/t$ and the distance l is given by the area enclosed by the graph, $l = (v \times t)/2$. Therefore,

$$\text{Work (KE)} = m \times \frac{v}{t} \times \frac{vt}{2} = \tfrac{1}{2}mv^2$$

Example 6.4 A car having a mass of 1000 kg has velocity of 15 m/s. Find the kinetic energy possessed by the moving car.

$$W = \tfrac{1}{2}mv^2 = \tfrac{1}{2} \times 1000 \times 15^2 = 112\,500 \text{ J}$$

Question 6.4 If the velocity of the car in Example 6.4 is doubled, find the new value of kinetic energy possessed by the car.   (450 000 J)

When the car is braked to a stop this energy is dissipated in the form of heat. As the brake shoe can only dissipate a given amount of energy in a given period of time, it is easy to see, by comparing the answers to Example 6.4 and Question 6.4, why the braking distance will increase with the square of the car's velocity.

The pendulum

When a pendulum is swinging, potential energy and kinetic energy are constantly being interchanged.

When the pendulum mass is raised to position 1 in Fig. 6.3, it has

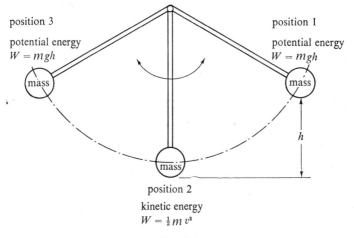

Fig. 6.3

ENERGY

been given the potential energy Mgh joules. When the pendulum is released this energy will cause it to accelerate, and as its velocity increases the potential energy begins to change into kinetic energy.

In position 2, the lowest point of swing, the potential energy is zero as $h = 0$. All the energy is now kinetic, $\frac{1}{2}Mv^2$ joules.

Once the pendulum has passed position 2 the kinetic energy will decrease and the potential energy will increase until, at position 3, the energy is again all potential.

Ignoring the effects of friction and windage on the pendulum:

$$\text{Energy position 1} = \text{Energy position 2} = \text{Energy position 3}$$
$$mgh = \tfrac{1}{2}mv^2 = mgh$$

At the pendulum's positions between 1 and 2, or 2 and 3, the sum of the potential and kinetic energies will be the potential energy at position 1 or 3 or the kinetic energy at position 2.

Example 6.5 A pendulum has a mass of 500 g and is lifted to a height of 25 cm. Find the potential energy given to the pendulum and its maximum velocity gained during the swing.

$$\text{Potential energy} = mgh = \frac{500}{1000} \times 9\cdot 81 \times \frac{25}{100}$$
$$= 1\cdot 226 \text{ J}$$

The pendulum will have a maximum velocity at the bottom of its swing and in this position $mgh = \tfrac{1}{2}mv^2$. Therefore,

$$1\cdot 226 = \tfrac{1}{2}mv^2$$
$$v^2 = \frac{2 \times 1\cdot 226}{m} = \frac{2 \times 1\cdot 226}{500/1000} = 4 \times 1\cdot 226 = 4\cdot 904$$

thus $\quad v = \sqrt{4\cdot 904} = 2\cdot 214 \text{ m/s}$

Question 6.5 Find the potential energy, kinetic energy, and velocity of the pendulum in Example 6.5 when its height has fallen to 12·5 cm.

(0·613 J; 0·613 J; 1·564 m/s)

Work done on an inclined plane

Example 6.6 A mass of 4 kg is pulled up an inclined plane of slope 1 in 15 (a rise of 1 in a slope distance of 15) for a distance of 3 m (Fig. 6.4). If the friction force between the slope and the mass is 10 N, find (a) the work done in overcoming friction and (b) the total work done.

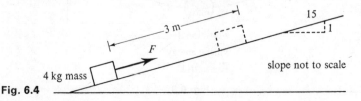

Fig. 6.4

(a) In a distance of 3 m the height gained is $\frac{3}{15} = 0.2$ m. The work done in lifting the mass vertically, that is the potential energy gained, is

$$mgh = 4 \times 9.81 \times 0.2 = 7.848 \text{ J}$$

The work done in sliding the mass against the friction force is

$$F \times l = 10 \times 3 = 30 \text{ J}$$

(b) Total work done is

$$7.848 + 30 = 37.848 \text{ J}$$

Question 6.6 A mass of 30 kg is allowed to slide down a slope of 1 in 10 (a fall of 1 in a slope distance of 10) for a distance of 2 m. If the friction force of the mass and the slope surface is 20 N, find (a) the potential energy released, (b) the work done in overcoming friction, (c) the kinetic energy gained by the mass after sliding 2 m, and (d) the velocity attained by the mass after sliding 2 m. (58·86 J; 40 J; 18·86 J; 1·125 m/s)

6.3 Power

The power rating of a machine will indicate how quickly work can be performed. For example, if a car engine is tuned to provide a greater output power, then the car will have greater acceleration and speed.

In definitive terms, *power is the rate of doing work.*

The symbol for power is P, and the unit is the watt (W)

$$\text{Power (watt)} = \frac{\text{Work (joule)}}{\text{Time (second)}} \quad \text{or} \quad P = \frac{W}{t}$$

The unit of power called the watt is equal to one joule per second (B.S. 3763).

Example 6.7 Find the power required to do 1000 J of work in a time of 20 s.

$$P = \frac{W}{t} = \frac{1000}{20} = 50 \text{ W}$$

Question 6.7 Find the power required to move a mass through a distance of 5 m in a time of 10 s if the force required to move the mass is 10 N. (5 W)

6.4 Efficiency

Efficiency is the ratio of the output work, which is the work done by a machine, to the input energy, which is the energy required by the machine to do the work. The symbol used for efficiency is η, which is the Greek letter eta.

$$\eta = \frac{\text{Output work}}{\text{Input energy}}$$

As the input energy must always be greater than the output work, the efficiency must always be less than 1. Thus the efficiency is said to be 'per unit' or part of 1. It is more usual to quote an efficiency as a percentage or part of one hundred.

$$\eta = \frac{\text{Output work} \times 100}{\text{Input energy}} \text{ per cent}$$

As $P = W/t$, and as most machines produce the output at the same time as the input, then

$$\eta = \frac{\text{Output power} \times 100}{\text{Input power}} \text{ per cent}$$

Example 6.8 The work done at the output of a machine is 10 000 J. If the efficiency of the machine is 60 per cent, find the energy required by the input and the loss of energy within the machine.

$$\eta = \frac{\text{Output work} \times 100}{\text{Input energy}} \text{ per cent}$$

therefore, Input energy $= \dfrac{\text{Output work} \times 100}{\eta}$

$$= \frac{10\,000 \times 100}{60} = 16\,667 \text{ J}$$

As a rough check the input energy (16 667 J) can be seen to be greater than the output work (10 000 J).

Loss of energy $= 16\,667 - 10\,000 = 6667$ J

Question 6.8 If the input power of a machine is 50 000 W and its output power is 40 000 W, find (a) the efficiency of the machine and (b) the power loss within the machine. (80 per cent; 10 000 W)

6.5 Practical units of power

As the watt is rather small for many purposes, a larger unit called the **kilowatt** (kW) is often employed; this unit is equal to 1000 W.

The unit of power still in general use for many machines is the **horsepower** (hp), which is 745·7 W. This is often approximated to 746 W.

6.6 Practical unit of work

From the formula $W = P \times t$, work can be seen to be the product of power and time. If the practical unit of power, the kilowatt, is multiplied by the practical unit of time, the hour, the result will be a practical unit of work.

Power of 1 kW × Time of 1 hour = 1 kilowatt hour (kWh)

$$1 \text{ kWh} = 1000 \text{ (W)} \times 60 \times 60 \text{ (s)}$$
$$= 3\,600\,000 \text{ J}$$

Example 6.9 An electric motor has an output rating of 5 hp. If the overall efficiency of the motor is 65 per cent, find the input to the motor in kilowatts.

$$5 \text{ hp} = 5 \times 746 = 3730 \text{ W}$$
$$\text{Input power} = \frac{3730 \times 100}{65} = 5740 \text{ W}$$
$$\text{or} = \frac{5740}{1000} = 5.74 \text{ kW}$$

Question 6.9 The output of an electric motor is rated as 10 hp and the input power required is 12·5 kW. Find the overall efficiency of the motor.

(59·8 per cent)

6.7 The cost of energy

Electrical energy is sold per kilowatt hour and for this purpose the kilowatt hour is called a *unit of electrical energy*.

Example 6.10 Find the cost of running the motor of Example 6.9 for a time of 8 hours at a cost of 1·5p per kilowatt hour.

$$\text{Input power} = 5.74 \text{ kW}$$
$$\text{Energy required} = 5.74 \times 8 = 45.92 \text{ kWh}$$
therefore, $\quad \text{Cost} = 45.92 \times 1.5 = 68.88p$

Question 6.10 Find the cost of running the motor of Question 6.9 for a time of 4 hours at a cost of 1·5p per kilowatt hour. (75p)

6.8 Work done by a variable force

In practice the force required to do work is seldom constant. An example of this is the compression of a coil spring. Initially, to compress the spring from its free length only a small force is required, but as the spring is further compressed the amount of effort required increases.

If, for example, a spring is found to compress 1 cm when a load of 1 N is applied, then to compress the spring a further 5 cm (a total compression of 6 cm) an additional 5 N will be required—that is, a total force of 6 N.

If a graph is drawn of force against compression (see Fig. 6.5) the resulting shape will be a sloping straight line.

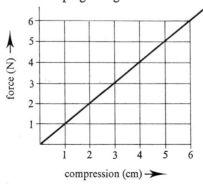

Fig. 6.5

The work done in compressing the spring will be the distance of compression multiplied by the *average* force during the compression.

If the spring is compressed from its free length then the average force will be:
$$\frac{\text{Final required force} - 0}{2}$$

Example 6.11 A spring is compressed from its free length of 5 cm to 4 cm by a force of 10 N. Find (a) the average force required and (b) the work done.

(a) Before compression   Force = 0
At end of compression   Force = 10 N
Thus,   Average force $= \dfrac{10 - 0}{2} = 5$ N

(b)   Work done = Force × Distance
$$= 5 \times \frac{5-4}{100} = 0.05 \text{ J}$$

The graph of force against compression is shown in Fig. 6.6. The area shown hatched in the diagram, calculated using the graph scales, will be:
$$\frac{\text{Base} \times \text{Height}}{2} = \frac{1/100 \times 10}{2} = 0.05$$

If this is compared with the answer to Example 6.11(b), it will be found to be the same. Therefore, *the scale area under the graph is the work done.*

Example 6.12 A spring requires a force of 12 N to cause an extension of 6 cm. Find the work done in extending the spring (a) from zero to 2 cm, and (b) from 2 cm to 6 cm.

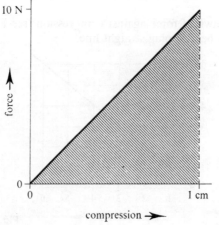

Fig. 6.6

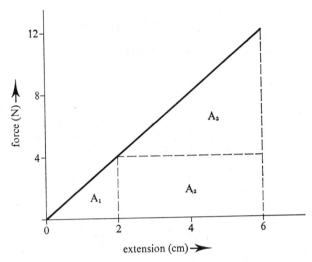

Fig. 6.7

Figure 6.7 shows the graph of force against spring extension.
(a) Work done in extending the spring from zero to 2 cm:

$$\text{Work done} = \text{Scale area } A_1 = \frac{2/100 \times 4}{2} = 0.04 \text{ J}$$

(b) Work done in extending the spring from 2 cm to 6 cm:

$$\text{Work done} = \text{Scale area } A_2 + \text{Scale area } A_3$$

$$\text{Scale area } A_2 = \frac{6-2}{100} \times 4 = \frac{16}{100} = 0.16 \text{ J}$$

$$\text{Scale area } A_3 = \frac{6-2}{100} \times (12-4) \times \tfrac{1}{2} = \frac{16}{100} = 0.16 \text{ J}$$

Therefore,
$$\text{Work done} = 0.16 + 0.16 = 0.32 \text{ J}$$

Question 6.11 A spring is extended from its free length of 40 cm to 60 cm by a force of 4 kgf. Find the work done in joules to extend the spring from 55 cm to 60 cm. (1·715 J)

Additional questions

6.12 One kilogramme of water is heated from 25°C to 75°C. Find the work done (a) in kilocalories and (b) in joules.

6.13 A mass of 1000 kg of water is pumped through a vertical height of 5 m. Find the work done by the pump on the water.

6.14 Explain the terms 'work' and 'power' and define one unit in which each may be measured.
   A pump having an efficiency of 85 per cent takes 30 min to raise 10 m³ of water through a vertical height of 20 m. Find the amount of work done on the water, the output power of the pump, and the amount of energy wasted in the process. The acceleration due to gravity may be taken as 9·8 m/s². (C.G.L.I.)

6.15 Explain what is meant by the term 'energy', defining one unit in which it may be measured.

State the principle of the conservation of energy.

(Based on C.G.L.I.)

6.16 What quantities are measured in newtons, joules, and watts? Give a definition of each of these three units.

A car of mass 1000 kg is driven at a steady speed of 45 km/h up a slope of 1 in 20 (i.e. rising 1 m in a slope distance of 20 m). Find the power used in overcoming gravity and the work done against gravity in driving the car a distance of 1 km. The acceleration due to gravity may be taken as 9·81 m/s². (C.G.L.I.)

6.17 A mass of 10 kg is pulled up a shaft by a rope 20 m long. The rope, which has a mass of 700 g/m, is completely wound in when the mass reaches the shaft top, and completely extended when the mass reaches the shaft bottom.

Draw a diagram showing how the force required varies as the mass is wound to the top of the shaft, then calculate (a) the work required to wind the mass (i) from the bottom to halfway up the shaft, (ii) the remainder of the way to the top, and (b) the total work done.

6.18 A body of mass 20 kg slides down an inclined plane from A to B as shown in Fig. 6.8. There is a constant friction force of 15 N in opposition

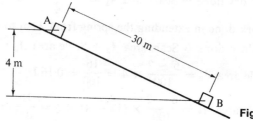

Fig. 6.8

to the motion. Find (a) the loss in potential energy, (b) the work done by gravity against friction, (c) the kinetic energy when the body reaches B, and (d) the percentage of the lost potential energy converted into heat.

(C.G.L.I.)

6.19 Explain the terms 'kinetic energy', 'potential energy', and 'power'. Name one unit of energy and one unit of power.

A water pump gives an output power of 1 kW while being used to raise 10 000 kg of water through a vertical distance of 20 m. Find (a) the time taken and (b) the input power to the pump, if the efficiency is 80 per cent. (C.G.L.I.)

6.20 Find the work done and the power used in each of the following cases:
  (a) a constant force of 20 N moves a body 5 m in the direction of the force in 2 s;
  (b) a constant couple of 10 N m turns a wheel through 30 complete revolutions in 1 min;
  (c) a machine with a 60 per cent efficiency is used to raise a mass of 100 kg through a vertical distance of 20 m in 5 min.

(Based on C.G.L.I.)

7 Machines

7.1 The machine
A machine may be defined as a device used to extend the work capabilities of man, either directly, or indirectly by extending the capabilities of another machine.

7.2 The lever
Figure 7.1 shows the basic essentials of the lever, which is perhaps the first, simplest, and still the most commonly used machine.

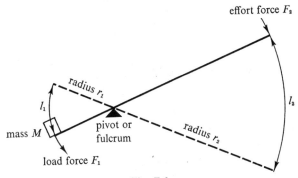

Fig. 7.1

The work done on the mass M is the product of the load force and distance moved, that is $F_1 \times l_1$. The energy required to do this work is the product of the effort force and distance moved, that is $F_2 \times l_2$.

As the work done by the load is the machine's output, and the work done by the effort is the machine's input:

$$\text{Efficiency } \eta = \frac{\text{Output work}}{\text{Input work}} = \frac{F_1 \times l_1}{F_2 \times l_2}$$

The length of arc of a circle is proportional to its radius, and so

$$l_1 : r_1 :: l_2 : r_2$$

or $\quad \dfrac{l_1}{r_1} = \dfrac{l_2}{r_2} \quad$ and thus $\quad \dfrac{l_1}{l_2} = \dfrac{r_1}{r_2}$

Therefore $\quad \eta = \dfrac{F_1 \times r_1}{F_2 \times r_2}$

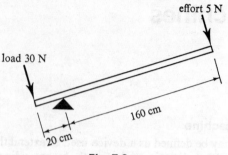

Fig. 7.2

7.3 Velocity ratio (VR)
The **velocity ratio** is

$$\frac{\text{Distance moved by the effort}}{\text{Distance moved by the load}}$$

or

$$\frac{\text{Velocity of the effort}}{\text{Velocity of the load}}$$

In the case of a lever this is either l_2/l_1 or r_2/r_1. For a lever this is often called the *leverage*.

7.4 Mechanical advantage (MA)
The **mechanical advantage** is the ratio

$$\frac{\text{Load}}{\text{Effort}}$$

For the lever shown in Fig. 7.1 the MA will be F_1/F_2. If this ratio is not greater than 1, then there would be no mechanical force advantage in the use of the machine.

7.5 Efficiency of a machine η

$$\eta = \frac{\text{Output work}}{\text{Input energy}} = \frac{\text{Distance moved by load} \times \text{Load}}{\text{Distance moved by effort} \times \text{Effort}}$$

$$= \frac{1}{\text{VR}} \times \text{MA} = \frac{\text{MA}}{\text{VR}}$$

Since the velocity ratio of a machine is normally fixed in its design, the mechanical advantage will vary with the machine's efficiency.

Example 7.1 The lever shown in Fig. 7.2 has the following dimensions: effort to fulcrum, 160 cm; load to fulcrum, 20 cm.

If an effort of 5 N is required to move a load of 30 N at a constant velocity through a distance of 5 cm, find (a) the velocity ratio, (b) the mechanical advantage, (c) the work done by the effort, (d) the work done on the load, and (e) the efficiency of the machine.

(a) $$VR = \frac{\text{Distance moved by the effort}}{\text{Distance moved by the load}}$$
$$= \frac{\text{Distance from fulcrum to effort}}{\text{Distance from fulcrum to load}}$$
$$= \frac{160}{20} = 8$$

(b) $$MA = \frac{\text{Load}}{\text{Effort}} = \frac{30}{5} = 6$$

(c) Work done by effort = Effort force × Distance moved
The distance moved = 5 cm × VR = 5 × 8 = 40 cm, therefore
$$\text{Work done} = 5 \times \frac{40}{100} = 2\ J$$

(d) Work done on load = Load force × Distance moved
$$= 30 \times \frac{5}{100} = 1.5\ J$$

(e) $$\text{Efficiency} = \frac{\text{Output work}}{\text{Input energy}} = \frac{1.5}{2} = 0.75 \text{ or } 75 \text{ per cent}$$

Alternatively,
$$\text{Efficiency} = \frac{MA}{VR} = \frac{6}{8} = 0.75 \text{ or } 75 \text{ per cent}$$

Question 7.1 A lever has a velocity ratio of 10 and a mechanical advantage of 7. If the overall length of the lever is 44 cm make a sketch of the lever showing the position of the fulcrum.

Find also (a) the effort required to lift a load of 2 kgf, (b) the work done (i) by the effort, (ii) on the load, if the load is moved a distance of 4 cm, and (c) the efficiency of the lever.

(0·285 kgf; 1·12 J; 0·785 J; 70 per cent)

7.6 Lever systems

Bell-cranked lever

The bell-cranked lever, shown in Fig. 7.3, is used to change the direction of application of a force as well as its value. The velocity ratio for the lever shown in the figure is l_2/l_1 and the direction of application of the load is at right angles to the direction of the effort.

Compound levers

Compound levers are a number of levers grouped into a common system. An example of a compound lever system is shown in Fig. 7.4.

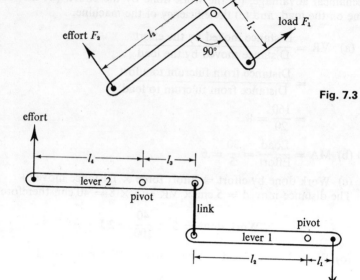

Fig. 7.3

Fig. 7.4

The velocity ratio of the first lever is l_2/l_1 and of the second lever is l_4/l_3. The velocity ratio of the compound lever system is the product of the two separate velocity ratios:

$$\text{VR} = \frac{l_2}{l_1} \times \frac{l_4}{l_3}$$

The length of the link between the two levers is of no importance to the calculation.

Example 7.2 If the dimensions of the lever system shown in Fig. 7.4 are l_1 4 cm, l_2 8 cm, l_3 3 cm, l_4 9 cm, and the overall efficiency is 80 per cent, find (a) the velocity ratio of the system and (b) the mechanical advantage of the system.

(a) $\quad \text{VR} = \dfrac{l_2}{l_1} \times \dfrac{l_4}{l_3} = \dfrac{8}{4} \times \dfrac{9}{3} = 2 \times 3 = 6$

(b) $\quad \eta = \dfrac{\text{MA}}{\text{VR}}$

therefore $\quad \text{MA} = \eta \times \text{VR} = \dfrac{80}{100} \times 6 = 4\cdot 8$

Question 7.2 Three similar levers, each having a velocity ratio of 3, are linked together in such a way that they obtain the maximum mechanical advantage. Sketch a suitable arrangement and state the velocity ratio of the lever system. (27)

7.7 The inclined plane

The velocity ratio for an object being moved up an inclined plane will be

$$VR = \frac{\text{Distance moved by the effort}}{\text{Distance moved by the load}}$$

From Fig. 7.5 this can be seen to be l/h.

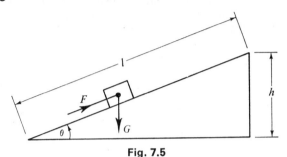

Fig. 7.5

If the slope angle is θ, then $\sin \theta = h/l$, and so

$$\frac{l}{h} = \frac{1}{\sin \theta} = \text{cosec } \theta$$

The mechanical advantage is

$$\frac{\text{Load}}{\text{Effort}} = \frac{G}{F}$$

and

$$\text{Efficiency} = \frac{\text{MA}}{\text{VR}} = \frac{G/F}{l/h} = \frac{Gh}{Fl} = \frac{G}{F} \sin \theta$$

Example 7.3 A barrel is rolled up a plank which has a slope angle of 30°. If the weight of the barrel is 40 kgf and the effort required to move the barrel at constant velocity is 25 kgf, find the efficiency of the arrangement.

$$\text{Efficiency} = \frac{G}{F} \sin \theta = \frac{40}{25} \times \sin 30$$

$$= \frac{40}{25} \times \frac{1}{2} = 80 \text{ per cent}$$

Question 7.3 Find the velocity ratio and mechanical advantage of the inclined plane described in Example 7.3. (2:1·6)

The effect of friction on the efficiency of the inclined plane

As the inclined plane becomes more horizontal the force between the surfaces will increase (see Section 4.9). As the friction force will thus also increase, the efficiency must decrease. Hence, the greater the velocity ratio, the worse the efficiency. This is a general condition for all machines and can be a limiting factor in their design.

A machine which uses the principle of the inclined plane to advantage is the screwjack. The revolution of the screw thread, which is merely a rotary inclined plane, will raise the load.

Example 7.4 The simple screwjack shown in Fig. 7.6 has a thread pitch of 5 mm and the thread is turned by a handle of 30 cm radius. Find the velocity ratio of the screwjack.

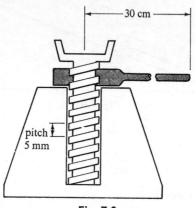

Fig. 7.6

If the handle is held at its extreme end and is turned through one revolution, the distance moved will be

$$2\pi r = 2\pi \times 30 = 188 \cdot 5 \text{ cm}$$

This is the distance moved by the effort. The load resting on the screwjack will be lifted a height of 5 mm or 0·5 cm, that is, the pitch of the thread.

The velocity ratio will be $188 \cdot 5/0 \cdot 5 = 377$.

Question 7.4 If the screwjack shown in Fig. 7.6 has an efficiency of 50 per cent, find (a) the mechanical advantage and (b) the effort required to raise a mass of 1000 kg. (188·5; 5·31 kgf)

7.8 Pulley blocks and tackles

The effect of a single pulley

If an effort is applied to raise a load (as shown in Fig. 7.7(a)), then for each 2 units of distance the rope is pulled up, the pulley—and therefore the load—will move up only one unit of distance (as shown by Fig. 7.7(b)), because each side of the pulley shares the total movement equally. This principle is used in most machines where ropes or cables are involved, and its effect should not be overlooked.

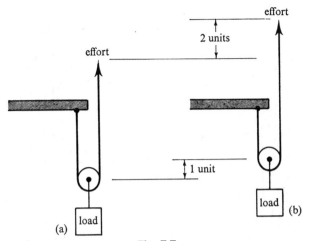

Fig. 7.7

Two pulleys

The velocity ratio of the pulley system shown in Fig. 7.8 is also 2. The function of the top pulley is to reverse the direction of the effort; it does not affect the velocity ratio.

In general, if the load and effort move in opposite directions, then the velocity ratio of the pulley system will be the number of pulleys, or the number of ropes, joining the pulley blocks.

Figure 7.9 shows a block and tackle with a total of five pulleys, the two blocks being joined by five ropes. To enable the ropes to be shown

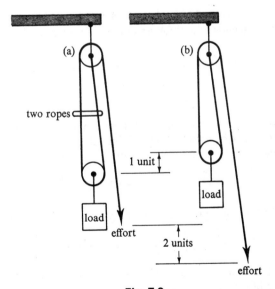

Fig. 7.8

PULLEY BLOCKS AND TACKLES **75**

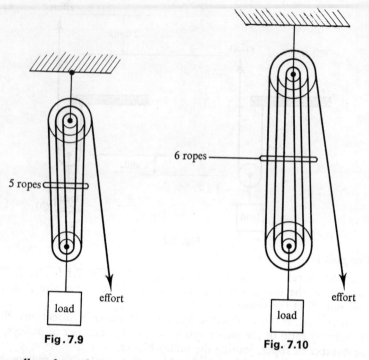

Fig. 7.9   Fig. 7.10

the pulleys have been drawn with different diameters. The velocity ratio of the machine will be 5.

Example 7.5   A mass of 10 kg is to be lifted by an effort of not more than 2·5 kgf by means of a block and tackle. Draw a workable arrangement.

The load will be 10 kgf and the mechanical advantage will be

$$\frac{\text{Load}}{\text{Effort}} = \frac{10}{2 \cdot 5} = 4$$

As the efficiency of the pulley system cannot be 100 per cent the velocity ratio must be greater than 4. If a velocity ratio of 5 is chosen, then the efficiency must be not less than

$$\frac{\text{MA}}{\text{VR}} = \frac{4}{5} = 0 \cdot 8 \text{ or } 80 \text{ per cent}$$

This might be too optimistic, so by trying a velocity ratio of 6,

$$\frac{\text{MA}}{\text{VR}} = \frac{4}{6} = 0 \cdot 67 \text{ or } 67 \text{ per cent}$$

which would seem more reasonable. Figure 7.10 shows the arrangement of the pulley blocks and ropes.

Therefore, if the efficiency is 67 per cent or better the effort required will not be greater than 2·5 kgf.

Question 7.5 A mass of 150 kg is to be lifted by an effort of not more than 75 kgf. Draw a suitable arrangement of pulleys if the minimum efficiency allowed is to be 50 per cent. State the velocity ratio of the machine. (4)

7.9 The differential axle

Figure 7.11 shows the arrangement of the machine.
The velocity ratio will be:

$$\text{VR} = \frac{\text{Distance moved by the effort}}{\text{Distance moved by the load}}$$

Let the axles turn through one revolution, then the effort rope will move a distance πd_2.

If rope A shortens a distance πd_2, rope B must pay out a distance πd_1, and the load will lift a distance $\frac{1}{2}(\pi d_2 - \pi d_1)$:

$$\text{VR} = \frac{\pi d_2}{\frac{1}{2}(\pi d_2 - \pi d_1)} = \frac{2d_2}{d_2 - d_1}$$

Since the effort rope will lengthen as the load is lifted, this machine is usually hand operated.

Example 7.6 A differential axle is used to lift an engine with a mass of 150 kg. If the axle diameters are 25 cm and 20 cm, find (a) the velocity

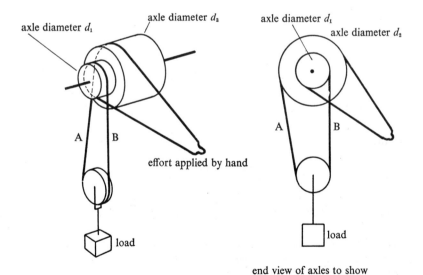

Fig. 7.11 Differential axle

ratio of the machine, and (b) its efficiency, if the effort required is 25 kgf.

(a) $$VR = \frac{2d_2}{d_2 - d_1} = \frac{2 \times 25}{25 - 20} = \frac{50}{5} = 10$$

(b) $$MA = \frac{Load}{Effort} = \frac{150}{25} = 6$$

therefore, $$Efficiency = \frac{MA}{VR} = \frac{6}{10} = 0.6 \text{ or } 60 \text{ per cent}$$

Question 7.6 A differential axle has axle diameters of 20 cm and 18 cm and is used to raise vertically a mass of 400 kg. If the overall efficiency is 65 per cent find (a) the velocity ratio and (b) the effort required. (20; 30·8 kgf)

7.10 The wheel and differential axle

Figure 7.12 shows the arrangement of a wheel and differential axle. The advantage of this machine is that it may be driven by a motor. The velocity ratio will be:

$$VR = \frac{\text{Distance moved by the effort}}{\text{Distance moved by the load}}$$

Let the wheel turn through one revolution, then the effort belt moves πd_3. The axles will also turn through one revolution and the load will lift or fall a distance $\frac{1}{2}(\pi d_2 - \pi d_1)$. Therefore

$$VR = \frac{\pi d_3}{\frac{1}{2}(\pi d_2 - \pi d_1)} = \frac{2d_3}{d_2 - d_1}$$

Example 7.7 A wheel and differential axle has the following dimensions: wheel, 8 cm diameter; axles, 4 cm and 3·8 cm diameter. A load of 160 kgf is raised by an effort in the belt of 4 kgf. Find (a) the velocity ratio, (b) the mechanical advantage, and (c) the efficiency of the machine.

(a) $$VR = \frac{2d_3}{d_2 - d_1} = \frac{2 \times 8}{4 - 3.8} = \frac{16}{0.2} = 80$$

(b) $$MA = \frac{Load}{Effort} = \frac{160}{4} = 40$$

(c) $$Efficiency = \frac{MA}{VR} = \frac{40}{80} = 0.5 = 50 \text{ per cent}$$

Question 7.7 In a wheel and differential axle, the wheel has a diameter of 20 cm and the axles have diameters of 6 cm and 5 cm respectively. Find (a) the machine's velocity ratio, (b) the mechanical advantage if a load of 200 kgf can be lifted by an effort of 8 kgf, and (c) the efficiency.
(40; 25; 62·5 per cent)

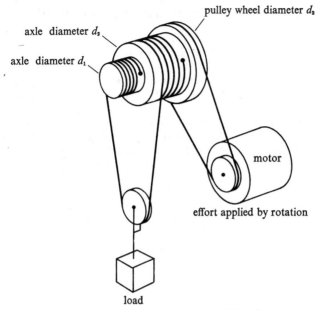

Fig. 7.12 Wheel and differential axle

7.11 Gearing

Figure 7.13 shows two gear wheels meshed. The gear wheel causing the rotation, the 'driver' gear wheel, provides the effort; the gear wheel being rotated, the 'driven' gear wheel, provides the load. For the gear wheels shown in the figure the velocity ratio will be

$$VR = \frac{\text{Distance moved by effort}}{\text{Distance moved by load}}$$

$$= \frac{\text{Number of revolutions of driven wheel}}{\text{Number of revolutions of driver wheel}}$$

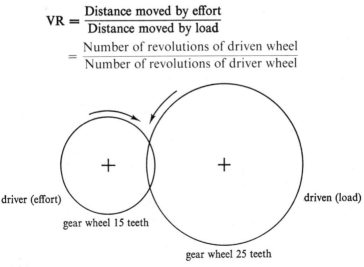

Fig. 7.13

If the driver gear wheel rotates once, the driven wheel will rotate $\frac{15}{25}$ of a revolution.

$$VR = \frac{1}{15/25} = \frac{25}{15} = 1\cdot67$$

As the velocity ratio is also (velocity of effort)/(velocity of load), the gears will have reduced the load shaft speed by a ratio of $1\cdot67:1$.

Example 7.8 An electric motor has a speed of 1400 revolutions per minute (rev/min) whilst the machine that it is to drive requires only a speed of 700 rev/min. Find (a) the velocity ratio of the gears required to reduce the speed and (b) the mechanical advantage if the efficiency of the gears is 80 per cent.

(a) $$VR = \frac{\text{Velocity of effort}}{\text{Velocity of load}} = \frac{1400}{700} = 2$$

(b) $$MA = VR \times \eta = 2 \times \frac{80}{100} = 1\cdot6$$

If the input and output driving shafts each have a radius of 1 cm, the force acting at a tangent to the output shaft will be 1·6 times the tangential force on the input shaft. This principle is shown in Fig. 7.14.

As torque is the product of tangential force and radius, then the output torque will be 1·6 times the input torque. A gearbox is thus often referred to as a 'torque convertor'.

It will be seen from Fig. 7.13 that the direction of revolution of the input and output shafts have been reversed, and if the direction is not to be reversed, an intermediate gear wheel (an 'idler' gear) must be used. The number of teeth on an idler gear wheel will not affect the velocity ratio.

As gearboxes can also be used to increase shaft speeds, the velocity ratio and the mechanical advantage can have a value less than unity.

An example of this is a car gearbox. Normally the gearbox provides either a speed reduction—that is, a torque increase—or a direct coupling. Some, however, also provide an overdrive which is a speed increase and therefore reduction in torque output.

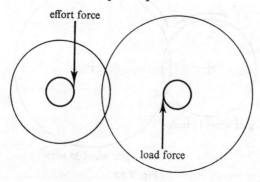

Fig. 7.14

80 MACHINES

Question 7.8 The gear train shown in Fig. 7.15 shows the driver gear G_1, with 10 teeth, driving a second gear wheel G_2, with 40 teeth. On the same shaft as G_2 a third gear G_3, with 10 teeth, is driving the driven gear wheel G_4, which has 50 teeth. The driving shaft has a speed of 3000 rev/min and its torque is 1·5 N m. If the efficiency of the gear train is 70 per cent, find (a) the velocity ratio, (b) the mechanical advantage, (c) the output torque, and (d) the output speed. (20; 14; 21 N m; 150 rev/min)

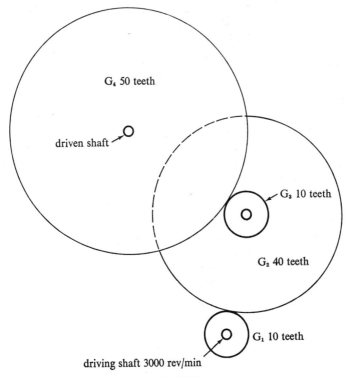

Fig. 7.15

7.12 The winch

Figure 7.16 shows the layout of a simple winch. In the figure, r is the radius of the winding handle to which the effort is applied; G_1 and G_2 are gear wheels which have N_1 and N_2 teeth, respectively; and d is the diameter of the winding shaft to which the load is applied. The velocity ratio of the machine will be:

$$VR = \frac{\text{Distance moved by the effort}}{\text{Distance moved by the load}}$$

Let the handle be turned one revolution, then the distance moved by the effort will be $2\pi r$.

THE WINCH   81

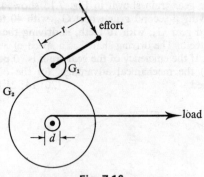

Fig. 7.16

The driven shaft will move $\pi d \times N_1/N_2$, therefore

$$VR = \frac{2\pi r}{\pi d(N_1/N_2)} = \frac{2rN_2}{dN_1}$$

If the effort is supplied by a motor and the pulley driving the winch has a diameter d_1, then the velocity ratio becomes $d_1 N_2/dN_1$.

Example 7.9 A winch has a driving pulley wheel of 50 cm diameter and a winding shaft of 5 cm diameter. The gears connecting the driving wheel and winding shaft have a velocity ratio of 4. If the load applied is 500 kgf and the overall efficiency is 70 per cent, find (a) the overall velocity ratio, (b) the mechanical advantage, and (c) the force in the belt rotating the driving pulley wheel.

(a) As the velocity ratio of the gears is 4, this must be the ratio N_2/N_1. Therefore the velocity ratio of the winch is

$$VR = \frac{d_1}{d} \times 4 = \frac{50}{5} \times 4 = 40$$

(b) $\quad MA = VR \times \text{Efficiency} = 40 \times \frac{70}{100} = 28$

(c) $\quad \text{Effort of belt drive} = \frac{\text{Load}}{MA} = \frac{500}{28} = 17.85 \text{ kgf}$

Question 7.9 A hand-driven winch is required to haul in a cable. The effort is to be limited to 10 kgf and it is estimated that the maximum force required will be 200 kgf. Allowing for an efficiency of 60 per cent, find (a) the mechanical advantage and (b) the velocity ratio.
Make a sketch of the winch and give typical dimensions for the crank and driving shaft, and details of gears required. (20; 33·3)

7.13 The worm and wheel

A worm and wheel gear is shown in Fig. 7.17. If the worm gear makes one revolution, then the worm thread will rotate the wheel a distance of one tooth. This statement assumes that the worm gear has a single-start thread; that is, one continuous thread.

The velocity ratio of a single-start worm and its wheel will be equal to the number of teeth on the wheel.

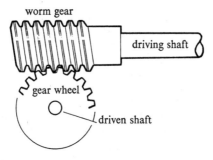

Fig. 7.17

Example 7.10 A single-start worm is meshed with a gear wheel of 50 teeth. Find (a) the velocity ratio and (b) the mechanical advantage if the efficiency is 75 per cent.

(a) VR = Number of teeth on gear wheel = 50
(b) MA = VR × Efficiency = 50 × 75/100 = 37·5

If a double-start worm gear had been used—that is, a worm gear which has two separate parallel threads—the velocity ratio would be halved, as one revolution of the worm would move the wheel by two teeth.

The driving and driven shafts of a worm and wheel cannot be reversed as the gears will lock in the reverse direction. This fact can be useful in, for example, a lifting device since the load will be automatically braked if the effort is released.

Question 7.10 The worm and wheel hoist shown in Fig. 7.18 is required to lift a load of 300 kgf through a distance of 4 m. The winding drum has a

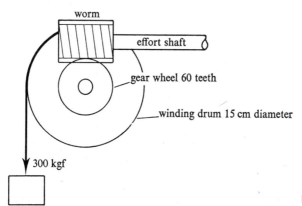

Fig. 7.18

THE WORM AND WHEEL 83

diameter of 15 cm, the wheel has 60 teeth, and the worm gear has a single start thread. If the efficiency is 75 per cent, find (a) the output torque in newton-metres, (b) the velocity ratio, (c) the mechanical advantage, and (d) the input torque in newton-metres. (220·5 N m; 60; 45; 4·9 N m)

Additional questions

7.11 A machine needs to be driven by a 100 rev/min ¼-hp motor but the only motor available has an output of 350 W, 4000 rev/min. What machine could be used to adapt the motor?
State its velocity ratio and any limitations. Calculate the minimum efficiency that could be tolerated.

7.12 A boat requires an effort of 1000 kgf to move it up a beach. If a winch is used which has a spindle of 15 cm diameter, a handle of 45 cm radius, and gear wheels of 75 and 15 teeth respectively, calculate the effort to be applied to the winch allowing for an efficiency of 60 per cent.

7.13 A load of 1000 kgf is to be raised 4 m by a force not exceeding 500 kgf. Sketch two machines that would be capable of this. State their speed ratio and find the effort if their efficiency is 75 per cent. Calculate the total energy used and the wasted energy.

7.14 A man wishes to raise a load of 150 kgf through a vertical height of 1 m and he is prepared to exert a force of not more than 30 kgf. Draw a diagram of a machine suitable for this purpose. State the speed ratio of the machine and calculate the minimum efficiency it would require. If the lifting process is carried out in 20 s, calculate the output power of the machine. (Based upon C.G.L.I.)

7.15 Explain the term efficiency as applied to a machine.
When using a lifting machine, it is found that an effort of 12·5 kgf must be exerted through a distance of 6 m in order to raise the load of 60 kgf through a distance of 1 m. Calculate the speed ratio, the mechanical advantage, and the efficiency of the machine. Sketch a machine to which the given data might refer. (Based upon C.G.L.I.)

7.16 Make a sketch of a pulley system which has a velocity ratio of 5. Calculate the efficiency of the system if an effort of 5 kgf is required to lift vertically a mass of 20 kg.

7.17 A force of 20 N is applied to the compound lever shown in Fig. 7.19. Find (a) the velocity ratio of the arrangement and (b) the value of the force F if the efficiency of the system is 90 per cent.

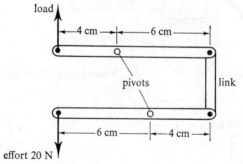

Fig. 7.19

7.18 (i) Explain an advantage of a worm and wheel gear when used in the mechanism of a hoist. (ii) A single-start worm gear is meshed with a 30-tooth gear wheel and a load torque of 200 N m is applied at the gear wheel. Find (a) the velocity ratio and (b) the input torque at the worm gear if the efficiency is 60 per cent.

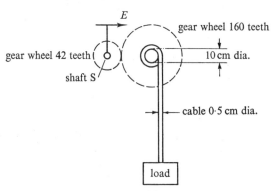

Fig. 7.20

7.19 Figure 7.20 shows details of a machine used for lifting a load. An effort E is applied by a handle on shaft S at a radius of 15 cm. Calculate (a) the effort E to raise a load of 0·5 tonnef, assuming the efficiency under these conditions to be 72 per cent, and (b) the torque input to the machine, stating units. (Based upon C.G.L.I.)

7.20 A screwjack has a single-start thread of pitch 0·5 cm. It is used to lift a load of 0·5 tonnef through 20 cm, the efficiency at this load being 55 per cent. Find (a) the number of turns made by the screw, (b) the work input, in joules, to the jack, and (c) the torque, in newton-metres, required to turn the screw. (Based upon C.G.L.I.)

8 Rotary Movement

8.1 The measurement of angles

An angle is a measurement of the distance that one line has rotated from another. One method of measurement is to divide the complete rotation into 360 parts and call each part a degree, as shown in Fig. 8.1.

Thus it is said that a circle contains 360°.

Another method is to rotate the line until the length of the arc formed by the moving end of the line is the same length as the line; that is, the arc formed is equal to the radius. This angle of rotation, shown in Fig. 8.2, is called a **radian**. The unit used for an angle measured in radians is 'rad'.

As the radius of a circle will divide into its circumference 2π times, there must be 2π radians in one complete rotation; that is, 2π rad = 360°.

Example 8.1 An angle is measured to be 60°. Express this angle in radians.

$$2\pi \text{ rad} : 360° :: \alpha \text{ rad} : 60°$$

$$\frac{2\pi}{360} = \frac{\alpha}{60}$$

thus, $\qquad \alpha = \dfrac{60 \times 2\pi}{360} = \dfrac{\pi}{3} = 1\cdot047 \text{ rad}$

Question 8.1 Express 45° in radians. $\qquad$ ($\pi/4$ or 0·785 rad)

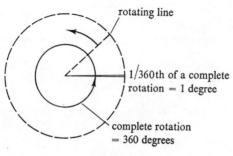

Fig. 8.1

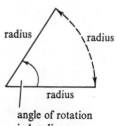

Fig. 8.2

8.2 Angular velocity

Angular velocity is the rate of rotation of a line, and is measured in radians per second (rad/s). The symbol used for angular velocity is ω, the Greek letter omega.

$$\omega = \frac{\text{Change of angle}}{\text{Time for the change}}$$

which may be expressed in the abbreviated form:

$$\omega = \frac{\Delta \text{ angle}}{\Delta t}$$

The rotation of a shaft is often expressed in revolutions per minute (rev/min, sometimes stated as r.p.m.).

Let N be the speed of rotation in revolutions per minute and n be the speed of rotation in revolutions per second (rotational frequency), then

$$n = \frac{N}{60}$$

The angle passed through per second will be $2\pi n$ radians and so $\omega = 2\pi n$ or $2\pi N/60$ rad/s.

Example 8.2 A motor has a speed of 3000 rev/min. What is the angular velocity in radians per second?

$$n = \frac{3000}{60} = 50 \text{ rev/s}$$

$$\omega = 2\pi n = 2\pi \times 50 = 314\cdot 2 \text{ rad/s}$$

Question 8.2 An electric motor has a speed of 1400 rev/min. Find the angular velocity in radians per second. (146·5 rad/s)

8.3 Peripheral velocity

The **peripheral velocity** is the speed of the rim of a wheel or the speed of a wheel's circumference.

From the definition of a radian we have found that when the angle at the centre of an arc is 1 rad, the length of the arc is equal to the radius. Therefore the length of arc will be the product of the angle, in radians, and the radius. This is shown in Fig. 8.3.

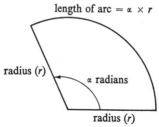

Fig. 8.3

The peripheral velocity will be the product of angular velocity and the radius:

$$\text{Peripheral velocity} = \omega r$$

If the angular velocity, ω, is measured in radians per second and the radius, r, in metres, then the peripheral velocity will be measured in metres per second (m/s).

Example 8.3 A wheel of diameter 6 cm has a speed of 1600 rev/min. Find (a) the angular velocity and (b) the pheripheral velocity.

(a) $$n = \frac{1600}{60} = 26 \cdot 66 \text{ rev/s}$$

thus, $$\omega = 2\pi n = 2\pi \times 26 \cdot 66 = 167 \cdot 6 \text{ rad/s}$$

(b) Peripheral velocity $= \omega r = 167 \cdot 6 \times \dfrac{6}{2} = 503$ cm/s or $5 \cdot 03$ m/s

Question 8.3 A pulley wheel has a mean diameter of 25 cm and is rotating at a speed of 1000 rev/min. Find the velocity of a belt fitted to the pulley.

(13·1 m/s)

8.4 Power

As power is the rate of doing work:

$$\text{Power (watts)} = \frac{\text{Work (joules)}}{\text{Time (seconds)}}$$

and work is the product of force and distance moved in the direction of the force:

$$\text{Work (joules)} = \text{Force (newtons)} \times \text{Distance (metres)}$$

then $$\text{Power} = \frac{\text{Force} \times \text{Distance}}{\text{Time}}$$

or $$\text{Power} = \text{Force} \times \text{Velocity}$$

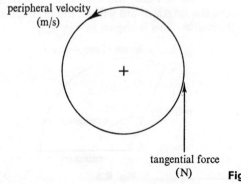

Fig. 8.4

The power of a rotating wheel will be the product of the peripheral velocity and the tangential force, as shown in Fig. 8.4.

Power = Tangential force × Peripheral velocity
      = Tangential force × Angular velocity × Radius

Torque is the product of tangential force and radius, therefore

Power (W) = Angular velocity (rad/s) × Torque (N m)
$$P = \omega T$$

Example 8.4 A force of 2 kgf is applied tangentially to a wheel of 10 cm diameter which is revolving at a speed of 1200 rev/min. Find the power developed.

$$n = \frac{N}{60} = \frac{1200}{60} = 20 \text{ rev/s}$$

$$\omega = 2\pi n = 2 \times \pi \times 20 = 125 \cdot 5 \text{ rad/s}$$

$$\text{Torque} = F \times r = 2 \times 9 \cdot 81 \times \frac{10}{2 \times 100} = 0 \cdot 981 \text{ N m}$$

therefore, $P = \omega T = 125 \cdot 5 \times 0 \cdot 981 = 123 \text{ W}$

Question 8.4 A motor delivers a power of ¼ hp to a wheel which has a speed of 800 rev/min. Find the torque of the wheel.   (2·23 N m)

8.5 Angular acceleration

Angular acceleration is the rate of increase of angular velocity. The symbol used is α (the Greek letter alpha), and its units are radians per second squared, rad/s².

$$\alpha = \frac{\text{Change of angular velocity}}{\text{Time for the change}}$$

which can be expressed in the abbreviated form $\Delta\omega/\Delta t$.

Example 8.5 A wheel rotating at 2100 rev/min is accelerated to 3000 rev/min in a time of 4 s. Find the average angular acceleration.

$$n_1 = \frac{2100}{60} = 35 \text{ rev/s}; \qquad n_2 = \frac{3000}{60} = 50 \text{ rev/s}$$

$$\omega_1 = 2\pi n_1 = 2\pi \times 35; \qquad \omega_2 = 2\pi n_2 = 2\pi \times 50$$

$$\alpha = \frac{\Delta\omega}{\Delta t} = \frac{\omega_2 - \omega_1}{\Delta t} \text{ (average)}$$

$$= \frac{2\pi(50 - 35)}{4} = \frac{30\pi}{4} = 23 \cdot 5 \text{ rad/s}^2$$

Question 8.5 A shaft is accelerated from 1800 rev/min to 2400 rev/min in a time of 3 s. Find (a) the angular acceleration and (b) the change in peripheral velocity if the diameter of the wheel is 10 cm.   (20·9 rad/s²; 3·142 m/s)

8.6 Inertia of a rotating body

In Chapter 2, Section 2.1, Newton's first law of motion was called the law of inertia. This law is equally true for a rotating wheel.

Figure 8.5 shows a wheel rotating with a constant angular velocity ω.

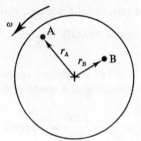

Fig. 8.5

The velocity of point A on the wheel will be $\omega \times r_A$ and that of point B, $\omega \times r_B$. As ω is constant, point A must be moving with a higher velocity than point B.

As the kinetic energy stored by the rotating wheel is proportional to the square of its velocity ($KE = \frac{1}{2}Mv^2$) point A must store a higher level of energy than point B.

8.7 The flywheel

A flywheel is used to smooth out a pulsating energy source; for example, the output of an internal combustion engine. This type of engine develops its energy from sudden gas expansions. The flywheel resists the sudden angular acceleration and carries the engine until the next gas expansion.

Examination of a flywheel will show that it is constructed with a heavy rim and a relatively light centre. This is because the kinetic energy of the rim is so much greater than that of its centre, and a flywheel thus constructed will be more effective.

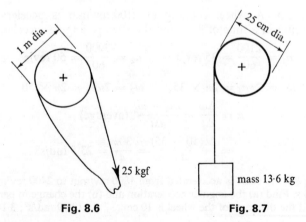

Fig. 8.6   Fig. 8.7

Additional questions

8.6 A man exerts a force of 25 kgf on an endless belt around a pulley of 1 m diameter, as shown in Fig. 8.6.
  (a) Find the torque of the pulley in newton-metres.
  (b) What power of motor could be used to drive the pulley at a speed of 10 rev/min?

8.7 A 1 hp motor turns a pulley wheel of 0·5 m diameter at 20 rev/min. Three types of belt are available which have breaking strengths of (a) 100 kgf, (b) 200 kgf, and (c) 300 kgf. Which of these belts would have the required strength?

8.8 A container of water is to be lifted up a shaft by a cable winding onto a drum of 25 cm diameter. If an electric motor turns the drum at a speed of 60 rev/min, and the container and water have a total mass of 13·6 kg, find the output torque and power of the motor (Fig. 8.7).

8.9 A gramophone turntable is revolving at a velocity of $33\frac{1}{3}$ rev/min. If the maximum and minimum radii at which the pick-up head operates are 14·5 cm and 6·5 cm respectively, find (a) the angular velocity and (b) the maximum and minimum velocities of the record passing the stylus.

8.10 A car travelling at a speed of 55 km/h accelerates to a speed of 75 km/h in a time of 30 s. If the car has wheels of 0·5 m diameter, find (a) the angular velocity before acceleration, (b) the angular velocity after acceleration, and (c) the angular acceleration.

9 Expansion of Solids and Liquids

9.1 Linear expansion of a solid

A **linear** expansion is an expansion of one dimension of a body. For example, as a railway line is very long in relation to its width and depth, its linear expansion is important since any expansion of its length could cause distortion of the line.

Coefficient of linear expansion

The increase in unit linear length per degree rise in the temperature of a material is called its **coefficient of linear expansion**.

The symbol used is α, which is the Greek letter alpha. The value of the coefficient of linear expansion is quoted at various temperatures dependent upon the material's normal use. For example, the coefficient for carbon steel is 11.4×10^{-6} per unit length per °C at 20°C.

The formula to find the length at any other temperature is

$$l_2 = l_1(1 + \alpha t)$$

where l_2 is the new length, l_1 is the length at the temperature at which the coefficient α is quoted, and t is the rise in temperature. If the temperature falls, t becomes a negative value.

As the value of the temperature coefficient is very small, only a small error is introduced if the initial temperature is not that at which the value of temperature coefficient is quoted. A correction is not normally necessary in practice unless extreme accuracy is required.

Example 9.1 A steel rod has a length of 1 m at a temperature of 20°C. Find its length at 50°C.

$$\begin{aligned} l_2 = l_1(1 + \alpha t) &= 1(1 + 11.4 \times 10^{-6}(50 - 20)) \\ &= 1(1 + 342 \times 10^{-6}) \\ &= 1 + 342 \times 10^{-6} \\ &= 1.000\,342 \text{ m} \end{aligned}$$

Question 9.1 Find the increase in length of the steel rod of Example 9.1 when its temperature is 100°C.   (0·912 mm)

9.2 Superficial expansion of a solid

The term **superficial** expansion denotes the expansion of two dimensions of the solid; that is, when the expansion of surface area is considered.

The coefficient of superficial expansion is denoted by the symbol β (the Greek letter beta), and is the increase in unit area per degree rise in temperature.

The formula used for finding the changed area is:

$$A_2 = A_1(1 + \beta t)$$

where A_2 is the new area, A_1 is the area at the temperature at which β is quoted, and t is the rise in temperature. If the temperature falls, t becomes a negative value. Normally the value of β is not obtained from tables but from the relationship $\beta = 2\alpha$.

Example 9.2 Find the increase in area of a sheet of steel having an initial area of 4 m², if its temperature is raised from 20°C to 40°C.

$$A_2 = A_1(1 + \beta t) = A_1(1 + 2\alpha t)$$
$$= 4(1 + 2 \times 11 \cdot 4 \times 10^{-6} \times 20)$$
$$= 4 + 1824 \times 10^{-6}$$

Increase in area $= A_2 - A_1 = 4 + 1824 \times 10^{-6} - 4$
$$= 1824 \times 10^{-6} \text{ m}^2$$
$$= 18 \cdot 24 \text{ cm}^2$$

Question 9.2 Find the area of the sheet of steel in Example 9.2 when its temperature is 90°C. (4·006 384 m²)

9.3 Cubical expansion of a solid or a liquid

Cubical expansion is the expansion of all three dimensions; that is, when the expansion of volume is considered.

Liquids have been added to this section. When a liquid expands its volume will increase, but individual dimensions will depend upon the container.

Coefficient of cubical expansion

The symbol for the coefficient of cubical expansion is γ, which is the Greek letter gamma. The value of γ is the increase in unit volume per degree rise in temperature, when the material has initially the temperature at which γ is quoted.

The formula used for the calculation of the changed volume is:

$$V_2 = V_1(1 + \gamma t)$$

where V_2 is the new volume, V_1 is the volume at the temperature at which the coefficient γ is quoted, and t is the rise in temperature. The value of γ is obtained by looking up the value of α in a table of coefficients of expansion and applying the formula $\gamma = 3\alpha$.

Example 9.3 A block of steel has a volume of 1000 cm³ at 20°C. Find its volume at 60°C.

$$V_2 = V_1(1 + \gamma t) = V_1(1 + 3\alpha t)$$
$$= 1000(1 + 3 \times 11\cdot 4 \times 10^{-6} \times 40)$$
$$= 1000(1 + 1368 \times 10^{-6})$$
$$= 1000(1\cdot 001\ 368) = 1001\cdot 368 \text{ cm}^3$$

Question 9.3 Find the decrease in the volume of the steel block of Example 9.3 when its temperature falls to 0°C. (0·684 cm³)

9.4 Practical applications

The thermometer

Figure 9.1 shows the principle of a thermometer. The glass bulb contains either mercury or alcohol, the only outlet from the bulb being the small-bore glass tube.

A slight increase in the volume of the liquid will cause a small cubical change in its volume but a large linear expansion along the small-bore glass tube.

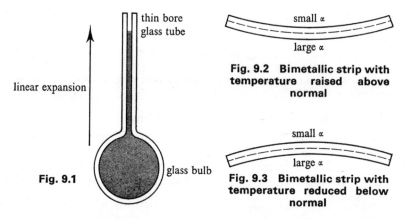

Fig. 9.1

Fig. 9.2 Bimetallic strip with temperature raised above normal

Fig. 9.3 Bimetallic strip with temperature reduced below normal

The amount of linear expansion is read against a scale calibrated in units of temperature.

The bimetallic strip

If two materials which have, respectively, a small and a large coefficient of linear expansion are joined together to form a bimetallic strip, then the strip will bend when it is heated or cooled, owing to one section expanding or contracting at a greater rate than the other. When heated the strip will bend as in Fig. 9.2, and when cooled it will bend as in Fig. 9.3.

Table 9.1 gives a list of materials, their coefficient of linear expansion, and the temperature at which this is quoted.

Table 9.1

Material	Temperature coefficient, per unit length per °C	Temperature, °C
Aluminium	$22 \cdot 5 \times 10^{-6}$	20
Brass	$18 \cdot 0 \times 10^{-6}$	24–100
Iron	$11 \cdot 9 \times 10^{-6}$	20
Mercury	$18 \cdot 2 \times 10^{-6}$	40
Steel	$11 \cdot 4 \times 10^{-6}$	20

Example 9.4 A bimetallic strip consists of a strip of brass and a strip of iron. If both strips are 5 cm long at 20°C, find the difference in lengths at 50°C.

Brass:
$$l_2 = l_1(1 + \alpha t) = l_1 + l_1 \alpha t$$

therefore the increase in length is

$$l_1 \alpha t = 5 \times 18 \times 10^{-6} \times 30 = 2700 \times 10^{-6} \text{ cm}$$

Iron: Increase in length is

$$l_1 \alpha t = 5 \times 11 \cdot 9 \times 10^{-6} \times 30 = 1782 \times 10^{-6} \text{ cm}$$

the brass section will be longer by

$$(2700 - 1782) \times 10^{-6} = 918 \times 10^{-6} \text{ cm or } 9 \cdot 18 \text{ μm}$$

Question 9.4 A brass and iron bimetallic strip has a length of 8 cm at 20°C. Find the lengths of the brass and iron at 80°C and the difference in their extensions. Draw a sketch to show how the bimetallic strip will bend.
(Brass, 8·008 65 cm; Iron, 8·0057 cm; 29·5 μm)

The thermostat

Figure 9.4 shows the principle of a simple thermostat, and this may be seen to be an application of the bimetallic strip.

In the figure the bimetallic strip is heated by an electric current

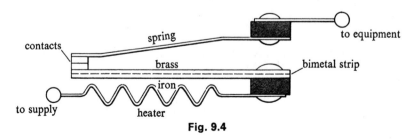

Fig. 9.4

passing through a resistance wire, bending the strip in a downward direction.

The electrical contact fixed to the bimetallic strip will also move downwards, followed by the contact fixed to the spring steel. When the spring has lost its tension the contacts will part and break the electrical circuit, but as the circuit will be remade when the bimetallic strip cools, a mechanical locking device may be provided to keep the contacts apart until the thermostatic switch is manually reset.

A bimetallic strip is used in the thermostatically controlled electric pressing iron, the heat in this case being provided by the iron element. The thermostat contacts will open at a predetermined temperature, and close when the heat energy stored by the iron has fallen below the required level.

Effect of heat on structures

All structures will expand with heat and this must be taken into consideration when they are designed. Metal bridges, for example, will often have a sliding joint to allow for expansion.

It is also important to consider the coefficients of different types of material used in a structure. If two materials with widely differing coefficients are used, structural distortion will occur.

Additional questions

9.5 A steel bridge is constructed with an overlapping sliding joint at one end. If the total length of the bridge is 100 m, estimate the difference in the amount of overlap when the temperature changes from 20°C to 35°C. Take the coefficient of linear expansion for steel as $11 \cdot 4 \times 10^{-6}$ per unit length per °C.

9.6 Two electrical contacts are separated by a gap of 2 cm. A piece of aluminium is inserted between the contacts in order to fill the space except for a 50 μm gap. At what temperature will the contacts be made if the initial temperature is 20°C. Take the coefficient of linear expansion for aluminium as $22 \cdot 5 \times 10^{-6}$ per unit length per °C.

9.7 An aluminium ball which has a diameter of 2 cm at 100°C is cooled to a temperature of 20°C. Find (a) its volume at 100°C, (b) its volume at 20°C, and (c) its diameter at 02°C. (The volume of a sphere may be calculated from the formula $V = \frac{4}{3}\pi r^3$.)

9.8 A rectangular-shaped steel tank has the following dimensions at 15°C: 1·25 m long, 0·75 m wide, 1 m deep. Find the volume of the tank at 30°C.

9.9 The inside diameter of the spherical bulb of a mercury thermometer is 0·5 cm and the inside diameter of the mercury column tube is 0·1 mm. At 40°C the length of the column of mercury is 10 cm. Estimate the length of the mercury column at 100°C if the expansion of the glass tube and bulb is ignored. Take the coefficient of linear expansion for mercury as 182×10^{-6} per unit length per °C.

9.10 Some electrical heating appliances are thermostatically controlled using a bimetallic strip. Sketch a bimetallic strip and explain briefly how it functions.   (C.G.L.I.)

10 The Gas Laws

10.1 Effect of heat on a gas

When a gas is heated it expands in a similar way to a solid or a liquid if it is permitted to do so. As a gas must be contained, the gas will only expand if the container walls are flexible. If the container is inflexible the gas cannot expand, but in trying to do so will exert a pressure on the walls of the container.

Thus for a gas, the temperature, volume, and pressure are interconnected.

Figures 10.1 and 10.2 show, respectively, a perfectly flexible container and a rigid container.

Gas temperatures and pressures must be calculated using absolute values; that is, the degree Kelvin for temperature and newtons per square metre absolute (pascal) or kilogrammes-force per square centimetre absolute for pressure.

10.2 Absolute unit of temperature

If for any perfect gas—that is, a gas which will follow the gas laws—a graph is plotted of volume against temperature for a condition of constant pressure (as in Fig. 10.1), then the graph will have a straight line. If the graph lines, plotted for various types of gases, are extended as shown in Fig. 10.3, then all the lines will meet and have zero volume at a temperature of $-273°C$.

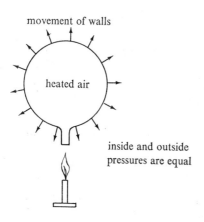

(Left) Fig. 10.1 Hot air balloon with very flexible walls

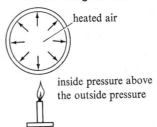

(Below) Fig. 10.2 Rigid container being heated

This temperature is thus chosen for the start of the absolute or Kelvin scale of temperature.

As a rise of 1 degree Kelvin is the same as a rise of 1 degree Celsius, then:

$$-273°C = 0 \text{ K}$$
$$0°C = 273 \text{ K}$$
$$100°C = 373 \text{ K}$$

which may be stated as K = °C + 273 or °C = K − 273.

Example 10.1 If a thermometer is indicating 20°C what is the absolute temperature?

$$\text{Absolute temperature K} = °C + 273$$
$$= 20 + 273 = 293 \text{ K}$$

Question 10.1 In the solution to a problem the absolute temperature was found to be 200 K. Find the temperature in degrees Celsius. (−73°C)

10.3 Absolute units of pressure

The normal pressure of the atmosphere is about 101 400 N/m² (or pa) above zero, and when a pressure gauge reads zero the actual or absolute pressure is therefore 101 400 N/m² (or pa).

In practice, a pressure gauge is more likely to be calibrated in kilogrammes-force per square centimetre, thus if the gauge reads zero, the absolute pressure in this type of unit will be 1·033 kgf/cm².

$$\text{Absolute pressure} = \text{Gauge pressure} + 1\cdot033 \text{ kgf/cm}^2$$

Example 10.2 A pressure gauge reads 15 kgf/cm². Find the absolute pressure.

$$\text{Absolute pressure} = 15 + 1\cdot033 = 16\cdot033 \text{ kgf/cm}^2$$

Question 10.2 If the absolute pressure is 274 000 N/m² (or pascals), find the reading that would be registered by pressure gauge if calibrated in newtons per square metre (or pascals). (171 000 N/m² or pa)

10.4 The gas laws

The constant pressure law (Charles's)

If the pressure of a given mass of gas remains constant, its volume is directly proportional to its absolute temperature.

$V \propto T$ when P is a constant and where $\propto$ means "is proportional to".

$$V_1 : T_1 :: V_2 : T_2$$
$$\frac{V_1}{T_1} = \frac{V_2}{T_2} \quad (P = \text{const.})$$

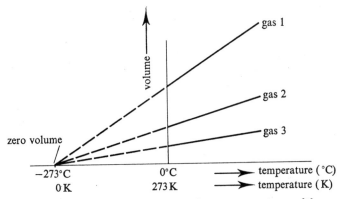

Fig. 10.3 Graph of volume against temperature with constant pressure

Example 10.3 A gas is expanded at a constant pressure from a volume of 1 cm³ to 2 cm³. If the temperature of the gas before expansion was 20°C find the final temperature.

$$\frac{V_1}{T_1} = \frac{V_2}{T_2} \quad (P = \text{const.})$$

therefore, $\quad T_2 = T_1 \dfrac{V_2}{V_1} = (20 + 273)\dfrac{2}{1} = 586 \text{ K}$

In degrees Celsius this is $586 - 273 = 313°C$.

Question 10.3 A gas is heated from 40°C to 200°C. If the initial volume was 0.2 m³, find the final volume if the pressure of the gas remains constant.

(0.303 m³)

The constant temperature law (Boyle's)

If the temperature of a given mass of gas remains constant, its volume will be inversely proportional to its pressure.

$$V \propto \frac{1}{P} \quad (T = \text{const.})$$

$$V_1 : \frac{1}{P_1} :: V_2 : \frac{1}{P_2}$$

$$\frac{V_1}{1/P_1} = \frac{V_2}{1/P_2}$$

therefore, $\quad P_1 V_1 = P_2 V_2 \quad (T = \text{const.})$

Example 10.4 A gas is expanded from a volume of 2 m³ to 5 m³ while its temperature remains constant. If, initially, its pressure was atmospheric, find its final pressure.

THE GAS LAWS   99

$$\text{Initial pressure} = 1.033 \text{ kgf/cm}^2$$
$$P_1 V_1 = P_2 V_2$$

therefore, $P_2 = \dfrac{P_1 V_1}{V_2} = \dfrac{1.033 \times 2}{5} = 0.4132 \text{ kgf/cm}^2$ absolute

or a gauge pressure of $0.4132 - 1.033 = -0.6198 \text{ kgf/cm}^2$, which is a vacuum of 0.6198 kgf/cm^2.

Question 10.4 The volume of a gas is changed from 1000 cm³ to 100 cm³, its temperature remaining constant. If the initial pressure was 1 kgf/cm² absolute, find the final pressure above atmospheric. (8·967 kgf/cm²)

The constant volume relationship

From Charles's and Boyle's laws it must follow that if the volume of a given mass of gas remains constant, then its pressure is directly proportional to the absolute temperature.

$$P \propto T \quad (V = \text{const.})$$
$$P_1 : T_1 :: P_2 : T_2$$
$$\frac{P_1}{T_1} = \frac{P_2}{T_2}$$

The characteristic equation of a gas

From Charles's law:
$$\frac{V_1}{T_1} = \frac{V_2}{T_2}$$

therefore, $\quad \dfrac{P_1 V_1}{T_1} = \dfrac{P_2 V_2}{T_2} \quad$ if $P_1 = P_2$

From Boyle's law:
$$P_1 V_1 = P_2 V_2$$

therefore, $\quad \dfrac{P_1 V_1}{T_1} = \dfrac{P_2 V_2}{T_2} \quad$ if $T_1 = T_2$

For the constant volume relationship
$$\frac{P_1}{T_1} = \frac{P_2}{T_2}$$

therefore, $\quad \dfrac{P_1 V_1}{T_1} = \dfrac{P_2 V_2}{T_2} \quad$ if $V_1 = V_2$

and in general $\quad \dfrac{P_1 V_1}{T_1} = \dfrac{P_2 V_2}{T_2}$

where neither P, V, nor T need be a constant, that is PV/T is a constant. This is called the characteristic equation of a gas.

In most practical conditions the pressure, volume, and temperature of a gas will vary simultaneously. When this happens the relationship $PV/T = $ constant remains true. If this relationship is remembered then both Charles's and Boyle's laws may be obtained as follows:

Bearing in mind the well-known initials P.C. and making them stand for constant **Pressure**, **C**harles's law, then from $P_1V_1/T_1 = P_2V_2/T_2$ and $P_1 = P_2$,

$$\frac{V_1}{T_1} = \frac{V_2}{T_2} \quad (P = \text{const.})$$

For Boyle's law the initials B.T. will stand for **B**oyle's law has constant **T**emperature, and from $P_1V_1/T_1 = P_2V_2/T_2$ where $T_1 = T_2$,

$$P_1V_1 = P_2V_2 \quad (T = \text{const.})$$

Example 10.5 A gas is compressed between gauge pressures of 2 kgf/cm² and 2·5 kgf/cm², its temperature rising from 19°C to 28°C. If the initial volume is 0·6 m³ find its final volume.

$$\frac{P_1V_1}{T_1} = \frac{P_2V_2}{T_2}$$

$P_1 = 2 + 1{\cdot}033 = 3{\cdot}033$ kgf/cm² absolute
$P_2 = 2{\cdot}5 + 1{\cdot}033 = 3{\cdot}533$ kgf/cm² absolute
$T_1 = 19 + 273 = 292$ K
$T_2 = 28 + 273 = 301$ K
$V_1 = 0{\cdot}6$ m³

therefore,

$$V_2 = \frac{P_1V_1T_2}{P_2T_1} = \frac{3{\cdot}033 \times 0{\cdot}6 \times 301}{3{\cdot}533 \times 292} = 0{\cdot}531 \text{ m}^3$$

Question 10.5 A gas is heated from a temperature of 17°C to 53°C. If the gas was initially at atmospheric pressure, find the final pressure gauge reading in kilogrammes-force per square centimetre if the volume of the gas is doubled. (*Hint:* $V_2 = 2V_1$.) (Vacuum of 0·453 kgf/cm²)

10.5 Heat as a form of energy

In Chapter 6, Section 6.2, it was shown that heat was a form of energy. Figure 10.4 is a graph showing the amount of heat required to evaporate a mass of 1 kg of ice, which initially has a temperature of 173 K.

Sensible heat

The line AB in Fig. 10.4 shows the temperature of the ice being increased from 173 K to 273 K, 211 kJ of energy being required for the process. This heat energy is termed 'sensible' as it causes a change in temperature. Thus the heating of the water from 273 K to 373 K, line CD, will also represent sensible heat.

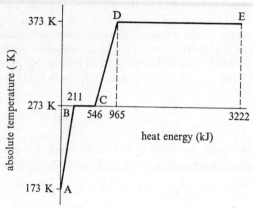

Fig. 10.4 Graph of temperature against heat energy for water

Sensible heat is denoted by the symbol h. The value of sensible heat for water may be calculated from:

h = Mass of water (kg) × Temperature rise (K) × 4187 J

Example 10.6 A 20 kg mass of water is heated from 20°C to 80°C. Find the sensible heat required.

$$h = \text{Mass} \times \text{Temperature rise} \times 4187 \text{ J}$$
$$= 20 \times 60 \times 4187$$
$$= 5\,024\,400 \text{ J}$$

Question 10.6 Find the sensible heat required to raise the temperature of 1 kg of water from 0°C to 100°C as shown by Fig. 10.4. (418 700 J)

Specific heat

Specific heat is the ratio of:

$$\frac{\text{Heat required to raise the temperature of a unit size mass one degree}}{\text{Heat required to raise the temperature of the same unit size mass of water one degree}}$$

The value obtained is a constant for a material, a selection of specific heat values being given in Table 10.1.

Water equivalent mass

All materials, other than water, will have an equivalent mass of water that will require the same quantity of heat energy to change its temperature:

Water equivalent mass = Mass of material × Specific heat

Table 10.1

Material	Specific heat	Material	Specific heat
Aluminium	0·214	Ice	0·504
Brass	0·094	Iron or steel	0·115
Copper	0·092	Lead	0·031
Glass	0·194	Zinc	0·095

Example 10.7 A metal of mass 1 kg requires 2000 J to raise its temperature 1°C. Find the specific heat of the metal and its water equivalent.

The heat required to raise the temperature of 1 kg of water 1°C = 4187 J. Therefore

$$\text{Specific heat} = \frac{2000}{4187} = 0\cdot478$$

$$\text{Water equivalent} = 0\cdot478 \times 1 = 0\cdot478 \text{ kg}$$

Question 10.7 A mass of 1 kg of ice is heated from $-100°C$ to $0°C$ as shown in Fig. 10.4. Find the sensible heat required if the specific heat of ice is 0·504. (211 125 J)

Latent heat

Section BC of Fig. 10.4 shows that 334 960 J of heat energy is required to melt the ice. The temperature remains constant at 273 K during this period, and the heat energy is therefore called 'latent'.

As the water is converted into steam (DE of Fig. 10.4), the temperature remains at 373 K. The amount of heat required is 2 257 000 J, and this heat is also latent.

The latent heat of the fusion of ice is thus 334 960 J/kg and the latent heat of the evaporation of the water is 2 257 000 J/kg.

Latent heat is denoted by the symbol L.

The effect of pressure on boiling point

Water will boil at a temperature of 373 K if the atmospheric pressure is 101 400 N/m² absolute or 1·033 kgf/cm² absolute. If the pressure is raised to 1·551 kgf/cm² the water will not boil until the temperature is about 395 K; that is, a rise of 12 K.

This principle is used in the water-cooling system of a car engine. The radiator cap is pressurized to about 1·32 kgf/cm² absolute and the water will not boil until a temperature of ~380 K is reached.

Question 10.8 Explain why a pressure cooker reduces the time factor in the cooking process.

Example 10.8 500 g of water at 20°C are converted into steam at atmospheric pressure. Find the quantity of heat required.

$$\text{Sensible heat } (h) = \text{Mass} \times \text{Temperature rise} \times 4187 \text{ J}$$
$$= \frac{500}{1000} \times (100 - 20) \times 4187$$
$$= 0{\cdot}5 \times 80 \times 4187$$
$$= 167\,480 \text{ J}$$
$$\text{Latent heat } (L) = 2\,257\,000 \text{ J/kg}$$
$$= 0{\cdot}5 \times 2\,257\,000 \text{ for } 0{\cdot}5 \text{ kg}$$
$$= 1\,128\,500 \text{ J}$$
$$\text{Total quantity of heat } (Q) = 167\,480 + 1\,128\,500$$
$$= 1\,295\,980 \text{ J}$$

Question 10.9 If the container of the 500 g of water in Example 10.8 has a mass of 600 g and is made from brass, find the extra heat required.

(18 850 J)

10.6 Heat transfer

Three methods exist by which heat may be transferred.

CONDUCTION: heat moving along a body.
CONVECTION: heat moving by the movement of hot gases or liquids.
RADIATION: heat moving through space.

An example of heat transfer commonly used in electrical engineering is the 'heat sink'.

Some electronic devices, semiconductors in particular, can be destroyed if they are overheated. The semiconductor can be bolted to a relatively large piece of metal called a heat sink, and the heat from the semiconductor will pass into the metal by conduction and will then be expelled from the large surface area of the metal by convection.

The heat sink should be so situated that air can circulate freely over its surface, but it should never be placed over another source of heat. The heat sink is often finned, rather like an air-cooled motor cycle cylinder, to assist the convection process.

Additional questions

10.10 A gas is expanded at constant temperature from a volume of 1 m³ to 2·5 m³. If the initial pressure was 4 kgf/cm² absolute, find the final pressure (a) in absolute units and (b) as a pressure gauge reading.

10.11 (a) State Charles's gas law.
   (b) A perfect gas is expanded at constant pressure from a volume of 10 cm³ to 50 cm³. When the gas has expanded to 20 cm³ the temperature is found to be 20°C. Find the temperature in degrees Celsius before and after expansion has taken place.

10.12 A hollow cylinder is heated to a temperature of 70°C and sealed so that the air inside is at atmospheric pressure. The cylinder is allowed to cool to 20°C. If the cylinder dimensions remain unchanged, find the pressure within the cylinder in (a) kilogrammes-force per square centimetre absolute and (b) newtons per square metre absolute.

10.13 A gas is heated from 10°C to 60°C by a contraction of its volume from 50 cm² to 20 cm². If the pressure of the gas before contraction was 1 kgf/cm² absolute, find the rise in its pressure.

10.14 A piston compresses a gas to one-fifth of its original volume during which time the pressure increases from 1·05 kgf/cm² absolute to 3·52 kgf/cm² absolute. If the initial temperature of the gas was 40°C, find the fall in temperature.

10.15 A body of mass 2 kg and specific heat 0·2 is heated from 20°C to 50°C. Find the heat energy gained.

10.16 A mass of 10 kg of ice at $-10°C$ is heated until the water formed has a temperature of 50°C. Find (a) the sensible heat required to raise the temperature of the ice from $-10°C$ to 0°C, (b) the latent heat required to convert the ice to water, (c) the sensible heat required to raise the temperature of the water from 0°C to 50°C, and (d) the total energy required. Draw to scale a graph of the temperature change against energy required.

10.17 Find the cost of heating 5 kg of aluminium from 17°C to 400°C if the cost of 1 kWh of energy is 2p.

10.18 If 0·5 m³ of air at 15°C enters an electric fan heater and leaves at 25°C, find the exit volume, assuming there has been no change in pressure.
(Based upon C.G.L.I.)

10.19 (a) Gas pressure of 6 kgf/cm² absolute acts downwards on the top of a piston 10 cm diameter, while normal atmospheric pressure acts upwards on the bottom. What is the nett downward force on the piston?
(b) The cylinder of an air compressor initially contains 500 cm³ of air at a pressure of 1 kgf/cm² absolute and temperature of 22°C. After compression the final volume is 100 cm³ and the final temperature is 80°C. Calculate the final pressure and express it as (i) an absolute pressure and (ii) a gauge pressure. Take normal atmospheric pressure to be 1·033 kgf/cm². (Based upon C.G.L.I.)

10.20 A volume V ft³ of air at 12°C is to be heated at constant pressure until its volume has increased by 20 per cent. To what temperature must it be heated? (C.G.L.I.)

10.21 Describe the methods by which a hot body can lose heat to its surroundings.
A component carrying a steady current is mounted on a heat sink. What practical details decide the temperature of the component?
(C.G.L.I.)

11 An Introduction to Electricity

11.1 The gold leaf electroscope

One of the earliest instruments used to demonstrate an effect of electricity was the electroscope. A version of this simple instrument is shown in Fig. 11.1. Gold leaf was used in the original instrument as it is possible to beat gold into extremely thin sheets or leaves.

If a glass rod is rubbed with a piece of silk and the rod is then applied to the protruding copper rod (Fig. 11.2), the foil leaves will part.

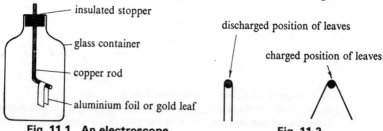

Fig. 11.1 An electroscope

Fig. 11.2

This happens because the foil has become electrically charged. The foil leaves can be discharged by touching the copper rod, when the leaves will revert to their original position.

To understand what happens when the leaves are charged it is first necessary to consider the basic construction of materials.

11.2 The construction of materials (properties of matter)

If a material is broken down into extremely small particles—smaller than can be seen by the most powerful microscope—a size of particle will be reached at which the material will entirely alter. For example, if a drop of pure water were divided into its smallest parts it would turn into hydrogen and oxygen gases.

Compounds and the molecule

Most materials are a mixture of other materials; that is, they are **compounds**. Pure water consists of two parts hydrogen and one part oxygen. Common salt consists of equal parts of sodium and chlorine.

The smallest part of a compound that can exist without change is called a molecule.

The smallest particle of pure water that can exist is called a **molecule**.

Elements and the atom

If a molecule of a compound is subdivided, the particles formed are called **elements**.

The smallest part of an element that can exist without change is called an atom.

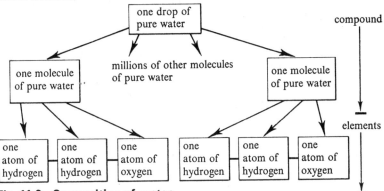

Fig. 11.3 Composition of water

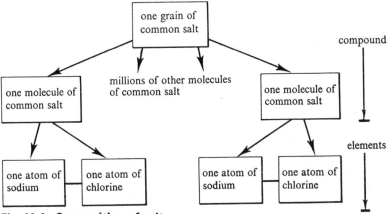

Fig. 11.4 Composition of salt

There are a number of materials, about ninety, that are constructed of only one type of **atom**. In addition to those already mentioned—hydrogen, oxygen, sodium, and chlorine—other examples are copper, aluminium, germanium, silicon, and arsenic. Thus, for example, a molecule of copper is also an atom of copper, since copper is an element.

The structure of the atom

The atom consists of a central section, called the **nucleus**, which is surrounded by particles causing an electric field.

The nucleus consists of minute masses called **protons** and **neutrons**. The proton is a mass which has an electrical charge of a positive nature whilst the neutron has no electrical charge at all.

The number of protons in the nucleus of an atom will decide the type of atom; for example, hydrogen has 1, oxygen has 8, carbon has 6, copper has 29, and gold has 79. The number of protons contained in the nucleus determines the **atomic number** of the element.

Around the nucleus revolve a number of much smaller, negatively charged particles called **electrons**. These are arranged in spherical layers called **shells** and there are normally an equal number of electrons and protons.

The simplest atom, hydrogen, will appear diagrammatically as shown in Fig. 11.5, while oxygen will appear as shown in Fig. 11.6.

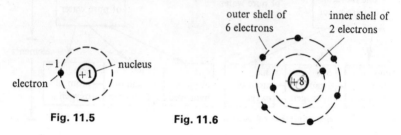

Fig. 11.5 **Fig. 11.6**

The mass of an electron is 0·00000000000000000000000000000904 kg (30 zeros), which is much more easily written as 9.04×10^{-31} kg. The mass of a proton is 1840 times greater, and is 1.663×10^{-27} kg.

11.3 The process of electrical charging

When a glass rod is rubbed by a piece of silk the heat energy produced by the friction will unbalance the electrical charges of the glass and the silk. The glass will lose electrons to the silk and will then be left in a state of positive charge. The silk, which now has an excess of electrons, will have a negative charge.

When the glass rod touches the copper rod of the electroscope, Fig. 11.1, electrons will be drawn from the aluminium foil via the copper rod to replace those lost by the glass. The positive charge between the plates will force the leaves apart.

If the electroscope terminal had been touched by the silk instead of the glass, the leaves would still part but the charge between the plates would be negative. These conditions are shown in Fig. 11.7.

If the electroscope terminal is touched by the silk, and if the glass rod is placed by the side of the electroscope as shown in Fig. 11.8, then the leaves will tilt towards the glass rod.

Thus the two leaves have been parted by a similar charge and attracted by an opposite charge. From these observations the relationship between charges can be seen to be:

Like charges repel; unlike charges attract

It is the second relation which binds or bonds together the atom and adjoining atoms in a structure.

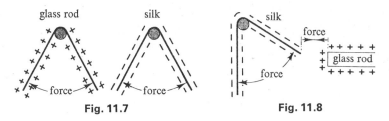

Fig. 11.7   Fig. 11.8

11.4 Electrical conductors

The most commonly used electrical conductors are copper and aluminium. These metals have electrons in the outer shell which are very mobile and are continually interchanging. Figure 11.9 shows the random progress of an electron in a conductor.

Fig. 11.9

If a positive charge is applied to one end of the conductor and a negative charge to the other end, the random movement turns into a one-way drift towards the positive charge.

11.5 Electrical insulators

Some insulation materials that are in common use for electrical cables are vulcanized rubber, polyvinyl chloride compound (P.V.C.), and compressed magnesium oxide. These materials have their outer electrons held in such a manner that they are immobile, and electric charge will not, therefore, be transmitted through the material under normal conditions.

11.6 Electric field

Between two oppositely charged bodies there will exist an **electric field**. As the electrons are being drawn towards the charged body having a lack of electrons, a physical force will exist between the bodies (Fig. 11.10).

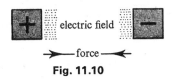

Fig. 11.10

In addition to a physical force between the bodies there must also exist an electrical force. This is called an **electromotive force** (e.m.f.).

11.7 Electromotive force (E or V)

The electromotive force is measured in **volts**.

Potential difference (*E* or *V*)

Potential difference is inseparable from electromotive force, and whenever an electromotive force exists between two bodies, there is also an equal potential difference.

Thus, the potential difference is also measured in volts.

Electromotive force of bodies

= Potential difference between bodies

Potential difference may be compared with the height gained by a mass (see Fig. 11.11). The difference in levels of the mass and the ground will give the mass the ability to do work (potential energy = Mgh joules).

In a similar way, the potential difference between the two charged bodies will give their charges the ability to do work.

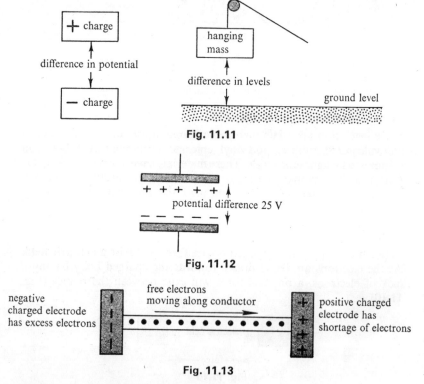

Fig. 11.11

Fig. 11.12

Fig. 11.13

11.8 Quantity of electric charge (*Q*)

As the electron has only a small electrical charge the SI unit used is the **coulomb** (C), which is much larger than the electron. One coulomb is the equivalent of $6 \cdot 28 \times 10^{18}$ electrons, and so the charge of one electron is $1 \cdot 58 \times 10^{-19}$ coulomb.

11.9 Capacity (*C*)

Capacity is the ability of a capacitor—which is two charged bodies, or electrodes, and the insulation between them—to allow a charge to be stored. The unit used is the **farad** (F).

When the charge held between two electrodes is Q coulombs and the potential difference between the electrodes is V volts, the capacity is given by the formula

$$C = \frac{Q}{V} \text{ farad}$$

The unit of capacitance, the farad, is defined in B.S. 3763 as follows:

The unit of electrical capacitance called the farad is the capacitance of a capacitor between the plates of which there appears a difference of potential of one volt when it is charged by a quantity of electricity equal to one coulomb.

Example 11.1 The charge held between the electrodes shown in Fig. 11.12 is 100 μC and the difference of potential between the electrodes is 25 V. Find the capacitance.

$$C = \frac{Q}{V} = \frac{100 \times 10^{-6}}{25} = 4 \times 10^{-6} \text{ F} \quad \text{or} \quad 4 \text{ μF}$$

The farad is an extremely large unit and values of capacitance normally range from about 1 pF (1×10^{-12} F) to 10 000 μF* (10 000 $\times 10^{-6}$ F).

The formula $C = Q/V$ is more often remembered in the transposed form $Q = CV$.

Question 11.1 If an e.m.f. of 200 V exists between two electrodes having a capacitance of 10 μF, find the charge stored. (2 mC)

11.10 Electric current (*I*)

Electric current is a measure of the rate of movement of electrons along an electrical conductor. An electrical conductor joining two electrically charged bodies is shown diagrammatically in Fig. 11.13.

At the positive electrode, which has had electrons removed in the charging process, free electrons from the conductor will replace electrons lost by the electrode.

In the conductor, the atoms which have lost their electrons will draw replacement electrons from adjacent atoms in the conductor. This process will be repeated along the conductor until the negatively charged electrode is reached, when the surplus electrons will restore the electron balance of the conductor.

This flow of electrons, or electric current, will continue until the electrodes are discharged.

* The larger capacitors could be rated in farads although, at present, this is not normal practice (10 000 μF = 0·01 F).

The greater the number of electrons moving along the conductor, the larger the electric current said to be flowing in the conductor.

Electric current is measured in **amperes** (A).

$$\text{Electric current} = \frac{\text{Quantity of electricity transferred}}{\text{Time taken}}$$

$$I \text{ (ampere)} = \frac{Q \text{ (coulomb)}}{t \text{ (second)}}$$

Example 11.2 If a quantity of 150 C is transferred in a time of 20 s find the average current during this time.

$$I = \frac{Q}{t} = \frac{150}{20} = 7 \cdot 5 \text{ A}$$

The formula $I = Q/t$ may be transposed to $Q = It$.

Question 11.2 If an electric current of 10 A flows through a conductor wire for a time of 15 s, how much electrical charge has been transferred?

(150 C)

The unit of electric current, the ampere, is defined electromagnetically (see Chapter 17, Section 17.8).

Using the formula $Q = It$, the coulomb may also be defined: *The unit of electric charge called the coulomb is the quantity of electricity transported in one second by a current of one ampere* (B.S. 3763).

Electron current flow and conventional current flow

From the explanation of electric current it is reasonable to indicate the direction of electric current, as in Fig. 11.14. If the direction of electron movement is to be shown, then this must be clearly indicated at the side of the arrow.

Before the reason for electric current was understood, the current was thought to be a flow from the positive electrode towards the negative electrode. It is still standard practice to follow this convention, as shown by Fig. 11.15.

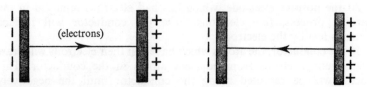

Fig. 11.14 Electron current   Fig. 11.15 Conventional current

11.11 Electrical resistance (*R*)

If conductor wires of different materials, but the same dimensions, are connected between electrodes which have the same potential difference, then the current in each wire will also be different.

Owing to their varying atomic construction, some materials will have electrons with greater freedom than other materials. Materials that have very free electrons are said to have a low resistance. Electrical resistance is measured in **ohms** (Ω).

A conductor must have a low resistance if it is to pass electrons easily, while an insulator must have an extremely high resistance if it is to stop electrons from flowing.

11.12 Semiconductors

A **semiconductor** material is one which has a resistance value between that of a conductor and that of an insulator. Its resistance will also decrease when it is heated (Fig. 11.16). Two basic semiconductor materials in common use are germanium and silicon.

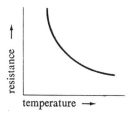

Fig. 11.16

When an electronic unit is said to employ semiconductors, the meaning goes much further than the simple definition given.

If different types of semiconductor material are joined, then the combination may be used for various purposes; for example, rectification, amplification, or switching. The two most common semiconductors are the semiconductor diode and the transistor.

11.13 Heat energy released by an electric current

As the removal of electrons from atoms will require energy, this energy —which is released on the breaking of the atomic bond—will cause the conductor to heat. This heat energy is proportional to the square of the electric current, the resistance of the conductor, and the time that the electric current is flowing.

$$W \text{ (joules)} = I^2 \text{ (ampere)} \times R \text{ (ohms)} \times t \text{ (seconds)}$$
$$W = I^2 R t$$

Example 11.3 A current of 2 A flows for a time of 10 min through a wire of resistance 70 Ω. Find the heat generated.

$$\begin{aligned} W &= I^2 R t \\ &= 2^2 \times 70 \times 10 \times 60 \\ &= 4 \times 42\,000 \\ &= 168\,000 \text{ J} \end{aligned}$$

Question 11.3 The heat output from a wire having a resistance of 10 Ω is 90 J/s. Find the current flowing in the wire. (3 A)

Additional questions

11.4 If a positively charged mass is brought near to a negatively charged mass, will the negative mass (a) try to move away from, (b) try to move towards, or (c) be unaffected by the positive mass?

11.5 (a) Make a diagrammatic sketch of an atom of oxygen.
(b) Explain why the electrons revolving around the nucleus are held to the nucleus.

11.6 Explain the essential difference between a compound and an element. Give two examples of each.

11.7 Name two materials that would be used in the manufacture of an electric cable. What are the essential properties of each material?

11.8 The charge stored by a capacitor of 8 μF is 400 μC. Find the potential difference between the electrodes of the capacitor.

11.9 A capacitor having a stored charge of 35 C is discharged by passing a current of 2 A between its terminals. Find the time taken to completely discharge the capacitor.

11.10 An electric lamp uses 7500 J of energy per minute. If the current taken by the lamp is 0·25 A, find the resistance of the lamp's filament.

11.11 A current of 2 A is passing through a resistance of 10 Ω. How long will it take to generate a heat energy of 800 J?

12 The Electrical Circuit

If two oppositely charged bodies are connected by an electrical conductor, then owing to their difference of potential, electrons will pass along the conductor (see Chapter 11, Fig. 11.14). This movement of electrons can only last until the opposite charges are equalized.

12.1 Electric circuit
When a negatively charged body is connected to a positively charged body, allowing an electric current to flow, then an electric circuit has been 'made'. An electric circuit is therefore a conducting path for an electric current.

A circuit diagram
A **circuit diagram** is a symbolic representation of the various components forming an electrical circuit.

12.2 Electric cell
An **electric cell** is usually a chemical device having two electrical connections or electrodes. The energy caused by reaction between chemicals causes electrons to be withdrawn from one electrode and passed to the other. The former becomes positively charged and the latter negatively charged. The action of an electric cell is dealt with more fully in Chapter 15.

When drawing a circuit diagram a cell is depicted by the symbol shown in Fig. 12.1. The polarity (the + and − signs) and the electromotive force (e.m.f.) may also be shown if desired.

Fig. 12.1   Fig. 12.2

Figure 12.1 shows a conductor wire, which has some resistance, joining the terminals of a cell.

Electrons will flow along the conductor causing it to heat: $W = I^2Rt$ joules (see Chapter 11, Section 11.13). The energy to cause this heat must be supplied from the chemical energy released from within the cell, and when the chemical energy is exhausted, the current will eventually weaken and cease.

12.3 The measurement of current

An instrument used to measure electric current is called an **ammeter**. To measure the current in a conductor the ammeter, which has a very low electrical resistance, must be inserted *into* the conductor as shown by Fig. 12.2.

12.4 The measurement of potential difference

An instrument used to measure the difference of potential between two electrodes is called a **voltmeter**. To measure the potential difference the voltmeter, which has a very high resistance, must be connected *across* the electrodes as shown by Fig. 12.3.

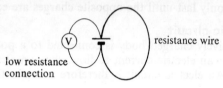

Fig. 12.3

12.5 Ohm's law

Georg Simon Ohm discovered that there was a definite relationship between the potential difference across the ends of a conductor and the magnitude of the current flowing in the conductor, and published the relationship in 1827. The law states:

$$\frac{\text{Potential difference across the conductor}}{\text{Current in the conductor}}$$

is a constant for a conductor material, assuming that the temperature is kept constant.

This fraction is called the conductor's **resistance**. The symbol used for resistance is R and the unit of measurement, the **ohm**, is shown by the Greek letter omega, Ω.

$$\text{Resistance} = \frac{\text{Potential difference across conductor}}{\text{Current in conductor}}$$

or $$R = \frac{E}{I} \qquad R = \frac{V}{I}$$

where R is the resistance in ohms (Ω), E or V is the potential difference (p.d.) in volts (V), and I is the current in amperes (A).

The unit of resistance

The following definition of the ohm as an SI unit is given in B.S. 3763:

The unit of electric resistance called the ohm is the resistance between two points of a conductor when a constant difference of potential of one volt, applied between these two points, produces in this conductor a current of one ampere, this conductor not being the source of any electromotive force.

The formula for Ohm's law is the most commonly used in electrical circuit calculations and should be remembered in its three forms:

$$R = \frac{E}{I}, \quad I = \frac{E}{R}, \quad E = IR$$

Example 12.1 A p.d. of 2 V causes a current of 0·5 A to flow in a conductor. Find the resistance of the conductor.

$$R = \frac{E}{I} = \frac{2}{0·5} = 4 \, \Omega$$

Question 12.1 Find the p.d. required to pass a current of 5 A through a resistance wire of 10 Ω. (50 V)

Question 12.2 A p.d. of 20 V is applied across a resistor wire of resistance 25 Ω. Find the current flowing in the wire. (0·8 A)

12.6 Power developed by a resistance

From the formulae $W = I^2Rt$ joules and $P = W/t$,

$$\text{Power} = \frac{I^2Rt}{t} = I^2R \text{ watts}$$

Applying the Ohm's law formula:

$$P = I^2R = \left(\frac{E}{R}\right)^2 R = \frac{E^2}{R} \text{ watts}$$

or

$$P = I^2R = I \times \frac{E}{R} \times R = IE \text{ watts}$$

Definition of the volt

From the formula $P = IE$ the unit of potential difference or electromotive force, the volt, is defined:

The unit of electric potential called the volt is the difference of potential between two points of a conducting wire carrying a constant current of one ampere, when the power dissipated between these points is equal to one watt (B.S. 3763).

Example 12.2 A current of 2 A flows through a resistance of 5 Ω. Find (a) the potential difference, (b) the power developed by the resistance, and (c) the work done by the current flowing for 5 min.

(a) $\quad E = IR = 2 \times 5 = 10 \text{ V}$
(b) $\quad P = I^2R = 2^2 \times 5 = 20 \text{ W}$

Alternatively, part (b) could be answered by

$$P = \frac{E^2}{R} = \frac{10^2}{5} = \frac{100}{5} = 20 \text{ W}$$

or

$\quad\quad P = IE = 2 \times 10 = 20 \text{ W}$

(c) $\quad W = Pt = 20 \times 5 \times 60 = 6000 \text{ J}$

Question 12.3 A p.d. of 6 V is applied across a resistance wire of 9 Ω. Find (a) the power developed, (b) the work done if the p.d. is maintained for a time of 15 min, and (c) the current in the wire.   (4 W, 3600 J, 0·667 A)

12.7 Resistors

It is often necessary to add to a circuit a value of resistance in order either to limit the current flowing in the circuit or to provide a potential difference. This can be done either by winding a length of resistance wire onto a former or by using a block of carbon. The symbol used for a resistor in a circuit diagram is shown in Fig. 12.4. This symbol is

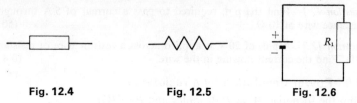

Fig. 12.4   Fig. 12.5   Fig. 12.6

internationally used and is preferred by B.S. 3939 to the symbol shown in Fig. 12.5, which is still in common use.

An alternative circuit to that shown in Fig. 12.1 is shown in Fig. 12.6.

All the circuit resistance is contained within resistor R_1. The conductor wires joining the resistor to the cell are of extremely low resistance and will not have a noticeable effect on the value of current in the circuit.

12.8 The series circuit

When resistors are joined as in Fig. 12.7, they are said to be connected in **series**. As there is only one conducting path an electric current can take, the value of the current in *all parts of the circuit* is the same.

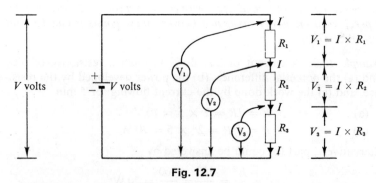

Fig. 12.7

Figure 12.7 also shows that the sum of the potential differences must equal the applied potential:

$$V = V_1 + V_2 + V_3$$

118 THE ELECTRICAL CIRCUIT

Applying Ohm's law to this relationship:

$$IR = IR_1 + IR_2 + IR_3 \quad \text{or} \quad R = R_1 + R_2 + R_3$$

Fall of potential along a series circuit

By drawing the resistors in a vertical line, as in Fig. 12.7, it is possible to see that voltmeter number 1 will read the supply potential, voltmeter number 2 will read the supply potential less the p.d. across resistor R_1, and voltmeter number 3 will read the supply potential less the p.d. of both R_1 and R_2.

The p.d. can therefore be said to fall along the circuit.

Summary of the rules of a series circuit

I is constant

$V = V_1 + V_2 + V_3 + V_4 + V_5$, etc.

$R = R_1 + R_2 + R_3 + R_4 + R_5$, etc.

Example 12.3 Four resistors, 10 Ω, 7 Ω, 5 Ω, and 2 Ω, are connected in series to a supply of potential 12 V. Find (a) the total resistance, (b) the current in the circuit, and (c) the p.d. of each resistor.

(a) $R = R_1 + R_2 + R_3 + R_4 = 10 + 7 + 5 + 2 = 24\ \Omega$

(b) $I = E/R = 12/24 = 0.5\ \text{A}$

(c) Potential difference across $R_1 = I \times R_1 = 0.5 \times 10 = 5\ \text{V}$
Potential difference across $R_2 = I \times R_2 = 0.5 \times 7 = 3.5\ \text{V}$
Potential difference across $R_3 = I \times R_3 = 0.5 \times 5 = 2.5\ \text{V}$
Potential difference across $R_4 = I \times R_4 = 0.5 \times 2 = 1\ \text{V}$

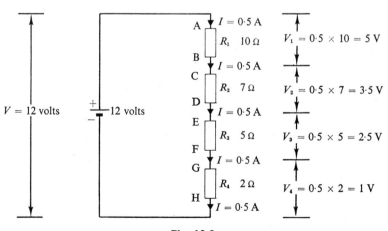

Fig. 12.8

Figure 12.8 shows that the result may be checked:

$$V = V_1 + V_2 + V_3 + V_4$$
$$12 = 5 + 3 \cdot 5 + 2 \cdot 5 + 1 \quad correct$$

Question 12.4 Three resistors, 5 Ω, 7 Ω, and 8 Ω, are connected in series. The current in each resistor is found to be 2 A. Draw a circuit diagram and find (a) the total resistance of the circuit, (b) the e.m.f. applied to the circuit, and (c) the p.d. across each resistor. (20 Ω; 40 V; 10 V, 14 V, 16 V)

Levels of potential

It is necessary to be able to state the level of potential of one part of a circuit with respect to another part of the same circuit.

Example 12.4 For the circuit diagram shown in Fig. 12.8 state the potential of points A and G with respect to point E.

A is positive to E by $5 + 3 \cdot 5 = 8 \cdot 5$ V

G is negative to E by $2 \cdot 5$ V

Question 12.5 For the circuit diagram shown in Fig. 12.8 state the potential of point D with respect to points B and H.
(D is negative to B by 3·5 V; D is positive to H by 3·5 V)

12.9 The parallel circuit

Resistors connected as shown by Fig. 12.9 are said to be connected in **parallel.**

Alternatively, this circuit could be drawn as shown in Fig. 12.10.

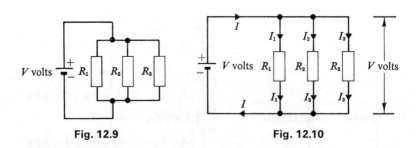

Fig. 12.9   Fig. 12.10

It is much easier to see from this figure that the *potential difference of each resistor is the same*. Each of the resistors will draw current from the cell.

$$\text{Current } I_1 = \frac{V}{R_1}; \quad I_2 = \frac{V}{R_2}; \quad I_3 = \frac{V}{R_3} \text{ ampere}$$

The total current (I) will be the sum of currents I_1, I_2, and I_3.

$$I = I_1 + I_2 + I_3$$

Applying Ohm's law to this relationship:

$$\frac{V}{R} = \frac{V}{R_1} + \frac{V}{R_2} + \frac{V}{R_3} \quad \text{or} \quad \frac{1}{R} = \frac{1}{R_1} + \frac{1}{R_2} + \frac{1}{R_3}$$

where R is the total resistance of the circuit.

Summary of the rules of a parallel circuit

V is constant

$$I = I_1 + I_2 + I_3 + I_4 + I_5, \text{ etc.}$$

$$\frac{1}{R} = \frac{1}{R_1} + \frac{1}{R_2} + \frac{1}{R_3} + \frac{1}{R_4} + \frac{1}{R_5}, \text{ etc.}$$

Example 12.5 Four resistors, 30 Ω, 40 Ω, 60 Ω, and 120 Ω, are connected in parallel to a 24 V e.m.f. (Fig. 12.11). Find (a) the total resistance, (b) the current in each resistor, and (c) the total current taken from the supply.

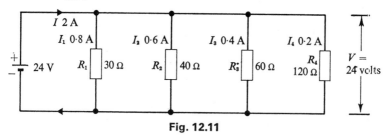

Fig. 12.11

(a) $\quad \dfrac{1}{R} = \dfrac{1}{R_1} + \dfrac{1}{R_2} + \dfrac{1}{R_3} + \dfrac{1}{R_4} = \dfrac{1}{30} + \dfrac{1}{40} + \dfrac{1}{60} + \dfrac{1}{120}$

$$= \frac{4 + 3 + 2 + 1}{120} = \frac{10}{120} = \frac{1}{12}$$

therefore, $\quad R = 12 \, \Omega$

(b) $\quad I_1 = \dfrac{V}{R_1} = \dfrac{24}{30} = 0.8 \text{ A}$

$\quad I_2 = \dfrac{V}{R_2} = \dfrac{24}{40} = 0.6 \text{ A}$

$$I_3 = \frac{V}{R_3} = \frac{24}{60} = 0.4 \text{ A}$$

$$I_4 = \frac{V}{R_4} = \frac{24}{120} = 0.2 \text{ A}$$

(c) $$I = \frac{V}{R} = \frac{24}{12} = 2 \text{ A}$$

Check: $2 = 0.8 + 0.6 + 0.4 + 0.2$ *correct*

The answers may be indicated on the circuit diagram, Fig. 12.11.

Question 12.6 Three resistors, 6 Ω, 9 Ω, and 18 Ω, are connected in parallel to a supply e.m.f. of 12 V. Find (a) the total resistance of the circuit, (b) the current in each resistor, and (c) the total current from the supply.

(3 Ω; 2 A, 1·33 A, 0·67 A; 4 A)

Two resistors in parallel

A circuit will often contain a pair of resistors connected in parallel. Using the formulae developed in Section 12.9, their equivalent resistance is:

$$\frac{1}{R} = \frac{1}{R_1} + \frac{1}{R_2} = \frac{R_2 + R_1}{R_1 \times R_2}$$

thus, $$R = \frac{R_1 \times R_2}{R_1 + R_2} \quad \left(\frac{\text{product}}{\text{sum}}\right)$$

Example 12.6 Using the above formula to recalculate the total resistance of the circuit of Example 12.5:

For the 30 Ω and 60 Ω resistors:

$$R = \frac{30 \times 60}{30 + 60} = \frac{1800}{90} = 20 \text{ Ω}$$

For the 40 Ω and 120 Ω resistors:

$$R = \frac{40 \times 120}{40 + 120} = \frac{4800}{160} = 30 \text{ Ω}$$

Now considering the 20 Ω and 30 Ω resistors:

$$R = \frac{20 \times 30}{20 + 30} = \frac{600}{50} = 12 \text{ Ω}$$

Question 12.7 Two resistors, 4 Ω and 6 Ω, are connected in parallel to a 6 V supply. Find the total resistance and supply current. (2·4 Ω, 2·5 A)

This formula has a great practical use in calculating the total resistance of two equal-valued resistors connected in parallel:

$$R = \frac{R_1 \times R_2}{R_1 + R_2}$$

If $R_1 = R_2$, then

$$R = \frac{R_1 \times R_1}{R_1 + R_1} = \frac{R_1 \times R_1}{2 \times R_1} = \frac{R_1}{2}$$

Thus the total resistance of a pair of equal-valued resistors, when connected in parallel, will be a half the value of each individual resistor.

Example 12.7 Two resistors, each having a value of 30 Ω, are connected in parallel to a 45 V supply. Calculate (a) the total resistance of the circuit, (b) the power dissipated by each resistor, and (c) the total power of the circuit.

(a) $\qquad\qquad\qquad$ Total resistance $= \dfrac{30}{2} = 15\ \Omega$

The current taken by each resistor is $V/R = 45/30 = 1\cdot5$ A, thus

(b) Power dissipated by each resistor is
$$I^2R = 1\cdot5^2 \times 30 = 67\cdot5\ \text{W}$$

(c) Total power dissipated by the circuit is
$$2 \times 67\cdot5 = 135\ \text{W}$$

If one resistor of 15 Ω had been used in this circuit, it would need a power rating in excess of 135 W. The two resistors of 30 Ω require power ratings which exceed 67·5 W.

One reason for the parallel connection of resistors is that smaller resistors which have a lower power dissipation may be readily available.

Question 12.8 (a) Two similar valued resistors are to be paralleled to produce a total resistance of 50 Ω. Find the value of resistors required.

(b) If a p.d. of 150 V is to be applied across the resistors, find the power dissipated by each resistor. $\qquad\qquad$ (100 Ω, 225 W)

12.10 The internal resistance of cells

Every source of e.m.f. must have some internal resistance. This may be presented diagrammatically, as shown in Fig. 12.12.

Figure 12.13 will be seen to be exactly the same electrically, as the internal resistance is in series with the external circuit resistance.

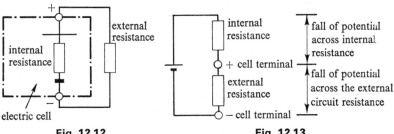

Fig. 12.12 $\qquad\qquad\qquad\qquad$ Fig. 12.13

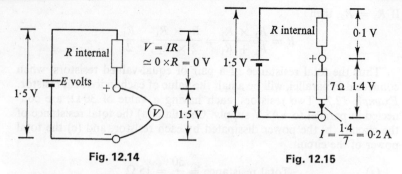

Fig. 12.14   Fig. 12.15

Example 12.8 A voltmeter, connected across the terminals of a cell, reads 1·5 V when the external circuit is disconnected (Fig. 12.14) and 1·4 V when a resistor of 7 Ω is connected across the cell terminals (Fig. 12.15). Find the internal resistance of the cell.

Since the voltmeter has a very high resistance, the current taken when only the voltmeter is connected to the cell is extremely small. Therefore the potential difference, $V = I \times R$, across the internal resistance is also extremely small. The reading on the voltmeter will thus represent the full e.m.f. of the cell, E volts.

When the 7 Ω resistor is connected to the cell, the p.d. across the resistor is 1·4 V. The difference between the two readings (that is, 1·5 − 1·4 = 0·1) must be the fall of potential across the internal resistance. This is as shown in Fig. 12.15.

The current in the 7 Ω resistor is $I = V/R = 1·4/7 = 0·2$ A. As this is a series circuit, the current must also be flowing in the internal resistance, and the value of this resistance is $R = V/I = 0·1/0·2 = 0·5$ Ω.

Question 12.9 A cell has a p.d. of 2 V when its load resistor is not connected. When measured with a load of 100 Ω the p.d. falls to 1·8 V. Find the internal resistance of the cell.   (11·11 Ω)

12.11 Cells connected in series

Figure 12.16 shows the connection of cells in series. If the cells are connected positive to negative, the total e.m.f. of the cells is the sum of the e.m.f. of each cell, and the total internal resistance is the sum of the internal resistance of each cell.

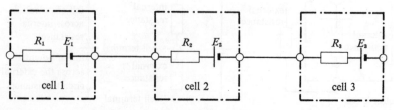

Fig. 12.16

Example 12.9 Three cells are connected in series across an external resistance of 10 Ω. If the e.m.f. of the cells are 1·5 V, 1·4 V, and 2·1 V, and the internal resistances are 0·1 Ω, 0·075 Ω, and 0·025 Ω respectively, find the current in the external resistance.

The total e.m.f. is

$$E = 1·5 + 1·4 + 2·1 = 5 \text{ V}$$

The total internal resistance is

$$R = 0·1 + 0·075 + 0·025 = 0·2 \text{ Ω}$$

The circuit can now be shown as in Fig. 12.17 and the total resistance of the whole circuit is

$$10 + 0·2 = 10·2 \text{ Ω}$$

The current in the circuit $= 5/10·2 = 0·49$ A, which is the current in the external resistor.

Question 12.10 Twelve cells, each having an e.m.f. of 1·5 V and an internal resistance of 0·02 Ω, are connected in series. If a resistance of 5 Ω is connected across the battery of cells, find the power dissipated in the external resistance.
(59 W)

The charging of cells in series

Example 12.10 Three batteries are to be charged from a 30 V direct current (d.c.) supply. If the e.m.f. and internal resistance of the batteries

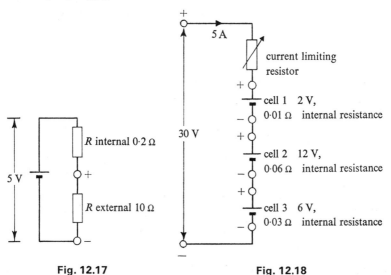

Fig. 12.17 **Fig. 12.18**

when being charged are 2 V, 0·01 Ω; 12 V, 0·06 Ω; and 6 V, 0·03 Ω respectively, show how the batteries should be connected. Calculate the value of any additional component that may be required when the charging current is to be limited to 5 A.

Figure 12.18 shows the suggested circuit. The current may be limited by a series-connected variable resistor.

The total battery e.m.f. is $2 + 12 + 6 = 20$ V.
The surplus p.d. is thus $30 - 20 = 10$ V.
To drop this p.d. at a current of 5 A the total circuit resistance must be

$$R = \frac{V}{I} = \frac{10}{5} = 2 \, \Omega$$

The total internal resistance of the batteries will be

$$0·01 + 0·06 + 0·03 = 0·1 \, \Omega$$

The additional value of resistance required in the circuit is therefore $2 - 0·1 = 1·9 \, \Omega$, which will be supplied by the variable resistance.

Question 12.11 Twenty similar secondary cells are to be connected in series and fed via a suitable resistor by an 80 V d.c. supply. The charging current is to be limited to 3 A and under these conditions the e.m.f. and internal resistance of each cell are known to be 2·5 V and 0·05 Ω respectively. Draw a circuit diagram showing the connection, and calculate the value of resistor required. (9 Ω)

12.12 Cells connected in parallel

Figure 12.19 shows a set of cells connected in parallel. This is an unusual method of connection except, perhaps, for the simultaneous charging of several batteries of a type dealt with in Chapter 15.

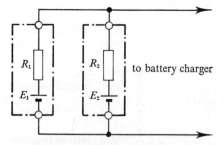

Fig. 12.19 The charging of parallel-connected cells

Care should be taken in the connection of cells in parallel, since even the smallest difference in the e.m.f. and internal resistance of each battery will cause the battery with the larger e.m.f. to discharge into the battery with the smaller e.m.f.

12.13 A series–parallel circuit

Example 12.11 Two resistors of 6 Ω and 12 Ω resistance are connected in parallel. A 2 Ω resistor is then joined in series to the combination, as shown in Fig. 12.20. If a 24 V supply of negligible internal resistance is connected across the circuit, find the current in each part of the circuit.

The total resistance of the parallel section is

$$\frac{6 \times 12}{6 + 12} = \frac{72}{18} = 4 \, \Omega$$

The total resistance of the circuit is

$$4 + 2 = 6 \, \Omega$$

Therefore,

$$\text{Current taken from supply} = \frac{E}{R} = \frac{24}{6} = 4 \, \text{A}$$

which is the current in the 2 Ω resistor.

The p.d. across the 6 Ω and 12 Ω resistance may be found by considering the equivalent circuit shown in Fig. 12.21.

The p.d. across the equivalent 4 Ω resistance is

$$V = IR = 4 \times 4 = 16 \, \text{V}$$

Both the 6 Ω and 12 Ω resistors must therefore have a 16 V potential difference. Therefore,

$$\text{Current in the 6 Ω resistor} = \frac{16}{6} = 2 \cdot 67 \, \text{A}$$

and

$$\text{Current in the 12 Ω resistor} = \frac{16}{12} = 1 \cdot 33 \, \text{A}$$

As a check, these currents should add up to 4 A:

$$2 \cdot 67 + 1 \cdot 33 = 4 \quad correct$$

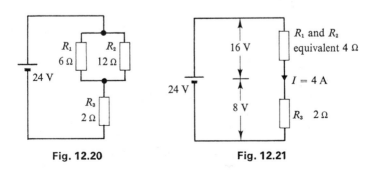

Fig. 12.20   Fig. 12.21

A SERIES-PARALLEL CIRCUIT

Question 12.12 Three resistors, 1 Ω, 1·5 Ω, and 3 Ω, are connected in parallel and this combination is connected in series to a 2·5 Ω resistor. If a 6 V supply is connected across the circuit, find (a) the total resistance of the circuit and (b) the current in each part of the circuit.

(3 Ω; 2 A, 1 A, 0·67 A, 0·33 A)

Example 12.12 A 24 V motor is supplied from fifteen 2 V cells each having an internal resistance of 0·001 Ω. When running, the motor is found to take a current of 10 A. To reduce the potential of the motor to 24 V a pair of equal valued parallel-connected resistors are placed in series with the motor (Fig. 12.22). Find (a) the value of each resistor,

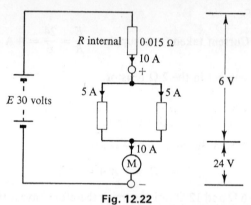

Fig. 12.22

(b) their minimum power rating, (c) the output power of the motor if the motor's efficiency is 60 per cent, and (d) the power lost by the internal resistance.

$$\text{Total } E \text{ of cells} = 15 \times 2 = 30 \text{ V}$$
$$\text{Total internal resistance} = 15 \times 0{\cdot}001 = 0{\cdot}015 \text{ Ω}$$
$$\text{Potential to be dropped} = 30 - 24 = 6 \text{ V}$$
$$R = \frac{V}{I} = \frac{6}{10} = 0{\cdot}6 \text{ Ω}$$

Thus, the external resistance required is

$$0{\cdot}6 - 0{\cdot}015 = 0{\cdot}585 \text{ Ω}$$

(a) Value of each resistor = $0{\cdot}585 \times 2 = 1{\cdot}17$ Ω
(b) $P = I^2R = 5^2 \times 1{\cdot}17 = 29{\cdot}25$ W (minimum)
(c) Input power to motor = $IV = 10 \times 24 = 240$ W, thus,

$$\text{Output power} = 240 \times \frac{60}{100} = 144 \text{ W}$$

(d) Power lost by the internal resistance of cells:

$$P = I^2R = 10^2 \times 0{\cdot}015 = 1{\cdot}5 \text{ W}$$

Additional questions

12.13 Three resistors, 2 Ω, 3 Ω, and 5 Ω, are connected in series to a 10 V supply. Find the current in the circuit and the p.d. of each resistor.

12.14 Two resistors, 2 Ω and 4 Ω, are connected in parallel. Find the value of the single resistor that could electrically replace them both.

12.15 Three resistors, 4 Ω, 6 Ω, and 2·4 Ω, are connected in parallel. Find (a) their equivalent resistance and (b) the current in each part of the circuit when the resistors are connected to a 6 V supply.

12.16 Four resistors are connected in series to a 12 V supply. If their values are 2 Ω, 10 Ω, 8 Ω, and 4 Ω, find the circuit current and the p.d. of each resistor.

12.17 Three resistors, 9 Ω, 12 Ω, and 18 Ω, are connected in parallel and an e.m.f. of 8 V is applied to the circuit. Find (a) the current in each part of the circuit and (b) the total resistance of the circuit.

12.18 Two resistors, 4·5 Ω and 9 Ω, are connected in parallel. A third resistor, 4·5 Ω, is then connected in series and an e.m.f. of 16 V and internal resistance 0·5 Ω is applied across the circuit. Find (a) the effective resistance of the circuit, (b) the current in each part of the circuit, and (c) the p.d. of each resistor.

12.19 The element of an electric fire is rated at 250 V, 0·5 kW. What is its resistance? (C.G.L.I.)

12.20 Resistors of 1 W, 2 W, and 5 W power rating are available for the construction of the network shown in Fig. 12.23. A p.d. of 160 V is to be maintained between A and B. Calculate the power dissipated in each resistor and draw a circuit diagram indicating suitable power ratings. (C.G.L.I.)

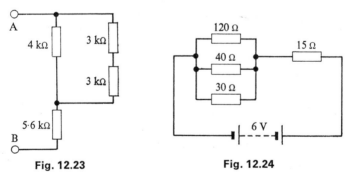

Fig. 12.23   Fig. 12.24

12.21 For the circuit shown in Fig. 12.24 calculate (a) the power dissipated in the 15 Ω resistor, (b) the current flowing in each of the resistors, and (c) the quantity of electricity supplied by the battery in 20 s.
What is meant by the statement that a conductor does not obey Ohm's law? (C.G.L.I.)

12.22 A d.c. motor with a series resistor of 10 Ω is used on a 240 V mains supply. When the motor is running the p.d. across the series resistor is found to be 40 V. Find (a) the p.d. across the motor and the current passing through it, (b) the power, in kilowatts, taken by the resistor and converted into heat, (c) the power, in kilowatts, taken from the mains supply, and (d) the horsepower developed if 55 per cent of the power input to the motor is converted to mechanical output.

(Based on C.G.L.I.)

12.23 The voltmeter shown in Fig. 12.25 reads 18 V when the switch is open and 16 V when the switch is closed. Find (a) the e.m.f. of the battery, (b) the current in the 4 Ω resistor, (c) the internal resistance of the battery, (d) the total quantity of energy converted into heat in the circuit in 30 s, (e) the power dissipated in the 4 Ω resistor, and (f) the short-circuit current of the battery. Why is it not necessary to consider the resistance of the voltmeter in your calculations? (C.G.L.I.)

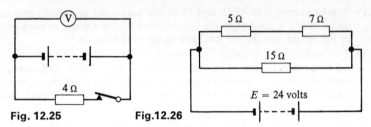

Fig. 12.25   Fig.12.26

12.24 An electric fire has an element of resistance 80 Ω and the supply is 200 V. Find (a) its power consumption and (b) the energy it will use in 6 hours. Name the unit in both cases. (C.G.L.I.)

12.25 Resistances are connected as shown in Fig. 12.26 to a battery that has an e.m.f. of 24 V and internal resistance of 0·5 Ω. Find (a) the current flowing through each resistor and (b) the p.d. across each resistor. (c) Draw a diagram showing how you would connect (i) a voltmeter to measure the p.d. across the 5 Ω resistor, and (ii) an ammeter to measure the current through the 5 Ω resistor. (C.G.L.I.)

12.26 What is meant by the statement that a conductor does not obey Ohm's law?

Draw Fig. 12.27, marking in the value and direction of the current at points A, B, C, and D. In an effort to find a fault in this circuit, the following voltmeter readings were taken. What was the fault?

Voltage between A and D = 6 V
Voltage between B and C = 3 V

(C.G.L.I.)

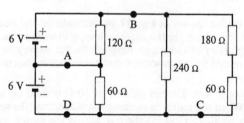

Fig. 12.27

130 THE ELECTRICAL CIRCUIT

13 Resistivity and the Effect of Temperature Change on Resistance

13.1 Resistivity

Effect of length on resistance

The resistance of a wire 20 cm long, having a resistance of 2 Ω for every 10 cm, will be 4 Ω. A length of a similar wire 60 cm long will have

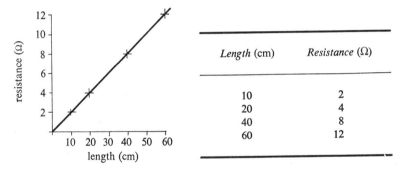

Fig. 13.1

a resistance of 12 Ω, that is, three times as much. Figure 13.1 shows the resistance for various lengths of the wire in tabulated and graphical form. As the graph has a straight line the resistance may be said to be proportional to the length of the wire:

$$R \propto l$$

Example 13.1 The resistance of 500 m of a two-wire cable is measured by joining the two wires at one end and measuring the resistance from the other end. If the measured resistance is 100 Ω find the resistance of 7 km of a similar cable.

$$R \propto l$$
$$R_1 : l_1 :: R_2 : l_2$$
$$\frac{R_1}{l_1} = \frac{R_2}{l_2}$$

thus,

$$R_2 = \frac{R_1 \times l_2}{l_1} = \frac{100 \times 7}{0\cdot 5} = 1400\ \Omega$$

This is the resistance of both wires in series, and is known as **loop** resistance.

Question 13.1 A resistor is to be made which has a resistance value of $2\cdot 5\ \text{k}\Omega$. If resistance wire is available which has a resistance of $400\ \Omega$ per 3 m, find the length of wire required. (18·75 m)

The effect of cross-section on resistance

If two similar wires, each of resistance $2\ \Omega$, are placed in parallel as shown by Fig. 13.2, then the resistance between terminals A and B will be:

$$\frac{1}{R} = \frac{1}{2} + \frac{1}{2} = \frac{2}{2}$$

therefore,

$$R = 1\ \Omega$$

Alternatively,

$$R = \frac{2 \times 2}{2 + 2} = \frac{4}{4} = 1\ \Omega$$

The area of an equivalent single wire would be doubled.

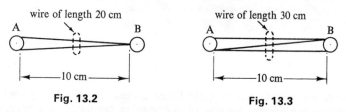

Fig. 13.2   Fig. 13.3

If three such wires are paralleled, as shown in Fig. 13.3, then the resistance A to B will be:

$$\frac{1}{R} = \frac{1}{2} + \frac{1}{2} + \frac{1}{2} = \frac{3}{2}$$

therefore,

$$R = \tfrac{2}{3}\ \Omega$$

and the area of an equivalent single wire would be trebled.

Figure 13.4 shows, in tabulated and graphical form, the resistance for various numbers of wires joined as Figs. 13.2 and 13.3. The area shown is the cross-sectional area of wire that would carry the current; that is, the product of the area of one wire and the number of wires.

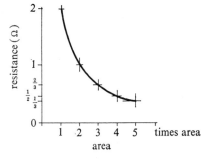

Fig. 13.4

At the left of Fig. 13.4 there is a graph of the resistance of a constant length of wire against its cross-sectional area. This graph has a law of the type $y \propto 1/x$ or, using the symbols R and A, $R \propto 1/A$.

Example 13.2 A wire of cross-sectional area 1 mm² has a resistance of 200 Ω. Find the resistance of a wire of the same length and material if its cross-sectional area is 1 cm².

There are 1000×1000 mm² in 1 m² and 100×100 cm² in 1 m², therefore there must be 10×10 mm² in 1 cm². The area is thus 100 times greater.

$$R \propto \frac{1}{A}$$

$$R_1 : \frac{1}{A_1} :: R_2 : \frac{1}{A_2}$$

$$\frac{R_1}{1/A_1} = \frac{R_2}{1/A_2}$$

$$R_1 \times A_1 = R_2 \times A_2$$

thus,

$$R_2 = \frac{R_1 \times A_1}{A_2} = \frac{200 \times 1}{100} = 2\ \Omega$$

Question 13.2 A wire of 0·5 mm diameter has a resistance of 1000 Ω. Find the resistance of the same length of a similar wire if its diameter is doubled.
(250 Ω)

If the length and the area of a wire are both variables, then $R \propto l/A$.

Example 13.3 A wire of length 10 m and cross-sectional area 2 mm² has a resistance of 50 Ω. If the wire is drawn out until its area is 1 mm², find the resistance of the wire.

RESISTIVITY

If the area of the wire is halved, its length must be doubled to 20 m.

$$R \propto \frac{l}{A}$$

$$R_1 : \frac{l_1}{A_1} :: R_2 : \frac{l_2}{A_2}$$

$$\frac{R_1 \times A_1}{l_1} = \frac{R_2 \times A_2}{l_2}$$

thus,

$$R_2 = \frac{R_1 \times A_1}{l_1} \times \frac{l_2}{A_2} = \frac{50 \times 2 \times 20}{10 \times 1} = 200 \: \Omega$$

Question 13.3 A resistance wire has a length of 20 m and a cross-sectional area of 4 mm². The wire is then drawn out until its length is increased four times. The resistance of the wire is now measured and found to be 160 Ω. Find the value of its resistance at the original dimensions. (10 Ω)

To enable the proportion $R \propto l/A$ to be changed into a formula a constant must be used:

$$R = \frac{l}{A} \times \text{constant}$$

The constant is called the **resistivity** of the material and is given the symbol ρ, which is the Greek letter rho:

$$R = \frac{\rho l}{A}$$

The value of resistivity is the resistance of a unit size cube of the material measured between opposite faces. A list of resistivity values for commonly used conductor materials is given in Table 13.1.

Table 13.1

Conductor material	Resistivity (Ωm at 0°C)
Aluminium	2.7×10^{-8}
Brass	7.2×10^{-8}
Carbon	4400×10^{-8} to 8600×10^{-8}
Constantan or Eureka	49×10^{-8}
Copper	1.59×10^{-8}
German silver	21×10^{-8}
Iron	9.1×10^{-8} *
Manganin	42×10^{-8}
Mercury	94×10^{-8}
Nickel	12.3×10^{-8} *
Tin	13.3×10^{-8}
Tungsten	5.35×10^{-8}
Zinc	5.57×10^{-8}

* Typical value.

The resistivity values of iron and nickel will vary according to construction and heat treatments.

The units shown in Table 13.1 are ohm-metres, which means that the size of cube used in the measurement was a metre-sided cube.

The units of resistivity are obtained by

$$R = \frac{\rho l}{A} \quad \text{therefore} \quad \rho = \frac{RA}{l}$$

$$\rho \text{ units} = \frac{\text{ohm} \times \text{metre} \times \text{metre}}{\text{metre}}$$

$$= \text{ohm metre}$$

The resistance of a centimetre-sided cube will be 100 times the resistance of a metre-sided cube, since although the length is 1/100th part of the metre, the area is 1/10 000th part of the metre, and as $\rho \propto A/l$

$$\rho \propto \frac{1/100}{1/10\,000} \propto \frac{10\,000}{100} \propto 100 \text{ times}$$

Thus the resistivity of copper in ohm-centimetres will be

$$100 \times 1{\cdot}59 \times 10^{-8} = 1{\cdot}59 \times 10^{-6}$$

Example 13.4 A coil is wound with a 4 m length of copper wire having a cross-sectional area of 1 mm². Find the resistance of the coil.

If the units of resistivity are ohm-metres, then both the length and the area of the wire must also be in metres:

$$\text{Length} = 4 \text{ m} \quad \text{and} \quad \text{Area} = 1 \text{ mm}^2$$

As there are 1000 × 1000 mm² in 1 m²:

$$1 \text{ mm}^2 = \frac{1}{1000 \times 1000} \quad \text{or} \quad 10^{-6} \text{ m}^2$$

Therefore,

$$R = \frac{\rho l}{A} = \frac{1{\cdot}59 \times 10^{-8} \times 4}{1 \times 10^{-6}} = 6{\cdot}36 \times 10^{-2} = 0{\cdot}0636 \, \Omega$$

Question 13.4 If the coil of Example 13.4 had been wound using constantan, find its resistance. (1·96 Ω)

13.2 Conductance

Conductance is the reciprocal of resistance. The symbol for conductance is G and it is measured in **siemens** (S):

$$G = \frac{1}{R}$$

Example 13.5 A resistor has a value of 2·4 Ω. Find its conductance.

$G = 1/R$. If $R = 2·4$ Ω,

$$G = \frac{1}{2·4} = 0·417 \text{ S}$$

Question 13.5 Two resistors have values of 4 Ω and 6 Ω. Find the value of their conductance. (0·25 S, 0·167 S)

Conductance may also be used when making calculations on parallel circuits.

Example 13.6 Three resistors, 2·4 Ω, 4 Ω, and 6 Ω, are connected in parallel. Find (a) their total conductance and (b) their total resistance.

From Example 13.5 and Question 13.5 the conductance of each resistor is 0·417 S, 0·25 S, and 0·167 S, respectively.

(a) As $1/R = 1/R_1 + 1/R_2 + 1/R_3$,

$$G = G_1 + G_2 + G_3 = 0·417 + 0·25 + 0·167$$
$$= 0·834 \text{ S}$$

(b) $R = \dfrac{1}{G} = \dfrac{1}{0·834} = 1·2$ Ω

Check: $\dfrac{1}{Rt} = \dfrac{1}{2·4} + \dfrac{1}{4} + \dfrac{1}{6} = \dfrac{5 + 3 + 2}{12}$ ∴ $Rt = 1·2$ Ω

Question 13.6 Two resistors, 3 Ω and 6 Ω, are connected in parallel. Find (a) their total conductance and (b) their total resistance. (0·5 S, 2 Ω)

13.3 Conductivity

Conductivity is the reciprocal of resistivity. The symbol for conductivity is σ (the Greek letter sigma) and it is measured in siemens per metre (S/m).

The value of conductivity will show how good a conductor the material will make; that is, the higher the value of conductivity the lower will be its resistivity and thus its resistance.

Example 13.7 From Table 13.1 the resistivity of copper is $1·59 \times 10^{-8}$ Ωm. Find the conductivity of copper.

$$\sigma = \frac{1}{\rho} = \frac{1}{1·59 \times 10^{-8}} = \frac{10^8}{1·59}$$
$$= \frac{100 \times 10^6}{1·59} = 62·9 \times 10^6 \text{ S/m}$$

Question 13.7 Using the value of resistivity given in Table 13.1, find the conductivity of aluminium. ($37·2 \times 10^6$ S/m)

13.4 Effect of temperature change on the value of resistance

If a graph is plotted of the resistance of a coil of copper wire against the temperature of the coil, it will assume the shape of the graph shown in Fig. 13.5. The temperature can be seen to increase by a straight line relationship.

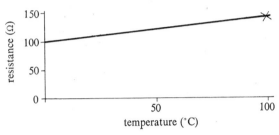

Fig. 13.5

If a copper wire of resistance 1 Ω at 0°C is heated to 1°C, its resistance will be 1·0043 Ω, an increase of 0·0043 Ω.

The change in the resistance of a 1 Ω resistor between 0 and 1°C is called the temperature coefficient of resistance for the resistor's material.

The symbol used for temperature coefficient is α, which is the Greek letter alpha.

If a copper wire having a resistance of 1 Ω is heated from 0°C to 100°C, then its change of resistance will be α × temperature rise or $\alpha\theta$, that is,

$$0{\cdot}0043 \times 100 = 0{\cdot}43 \; \Omega$$

The value of resistance will now be

$$1 + \alpha\theta = 1 + 0{\cdot}43 = 1{\cdot}43 \; \Omega$$

If the resistance of the coil had been 100 Ω at 0°C, its resistance at 100°C would be

$$R_0(1 + \alpha\theta) = 100 \times 1{\cdot}43 = 143 \; \Omega$$

where R_0 is the resistance at 0°C. These are the values shown plotted in Fig. 13.5.

The units of temperature coefficient are:

Resistance rise, for 1 ohm, from 0°C, for 1°C rise

or   ohm   per   ohm   at 0°C   per °C

Table 13.2 presents the values of temperature coefficient for some conductor materials.

Table 13.2

Conductor material	Temperature coefficient ohm/ohm at 0°C/°C
Aluminium	0·0038
Brass	0·001
Carbon	−0·0005
Constantan or Eureka	0·000 01 to −0·000 04
Copper	0·0043
German silver	0·000 27
Iron	0·0063
Manganin	0·000 025
Mercury	0·000 98
Nickel	0·0062
Tin	0·0044
Tungsten	0·0051
Zinc	0·0037

The worked example given above would normally be calculated as:

Example 13.8 A coil of copper wire has a resistance of 100 Ω when its temperature is 0°C. Find its resistance at 100°C.

$$R = R_0(1 + \alpha\theta) = 100(1 + 0.0043 \times 100)$$
$$= 100 \times 1.43 = 143 \, \Omega$$

Question 13.8 An aluminium cable has a conductor resistance of 25 Ω at a temperature of 20°C. Calculate its resistance at 0°C. (23·24 Ω)

13.5 The effect of positive temperature coefficient

Table 13.2 shows that nearly all conductor materials have a positive temperature coefficient; that is, their resistance will increase when they are heated.

A practical example of a positive temperature coefficient is the tungsten filament lamp. As the temperature of the filament of the lamp under working conditions will be about 2500°C, the resistance of the filament when cold will be very much lower than when operating.

Example 13.9 Estimate (a) the resistance of a tungsten filament lamp at 0°C and (b) the current that it will then pass if it has normally a power dissipation of 100 W with an applied potential of 250 V.

Assuming that the filament, under operational conditions, has a temperature of 2500°C and that the temperature coefficient for tungsten is 0·0051 ohm/ohm at 0°C/°C:

Resistance when hot
$$P = E^2/R$$
therefore
$$R = \frac{E^2}{P} = \frac{250^2}{100} = 625 \ \Omega$$

Current taken when hot $= \dfrac{250}{625} = 0\cdot 4$ A

(a) Resistance at 0°C
$$R = R_0(1 + \alpha\theta)$$
therefore,
$$R_0 = \frac{R}{1 + \alpha\theta}$$
$$= \frac{625}{1 + 0\cdot 0051 \times 2500}$$
$$= \frac{625}{1 + 12\cdot 75} = \frac{625}{13\cdot 75} = 45\cdot 5 \ \Omega$$

(b) Current taken $= \dfrac{250}{45\cdot 5} = 5\cdot 5$ A

Question 13.9 The resistance of the tungsten filament of a thermionic valve is 6 Ω when measured at 18°C. When the valve is operating, the filament takes a current of 0·15 A for an applied p.d. of 6·3 V. Estimate (a) the resistance of the filament at 0°C, (b) the current passing at 0°C, and (c) the temperature of the filament when operating normally.

(5·5 Ω, 1·14 A, 1300°C)

The answers to Example 13.9 and Question 13.9 show that the current taken when cold is very much greater than when hot. The filament of a lamp or valve is thus much more likely to fail when switched on than at any other time.

Superconduction
The answer to Question 13.8 shows that the resistance of an aluminium conductor will fall as its temperature is reduced. The principle of superconduction is that if the temperature of a conductor is reduced greatly its resistance will approach zero, as will its I^2Rt energy loss.

13.6 Alloyed resistance wire
In an alloy the temperature coefficient of some metals changes and can be made almost zero. Examples of such alloys are: manganin, which is 84 per cent copper, 12 per cent manganese, and 4 per cent nickel; and constantan or eureka, which is 60 per cent copper and 40 per cent nickel.

These alloys are suitable for wire wound resistors because their resistance value would be stable over a range of temperatures.

13.7 The effect of negative temperature coefficient

Carbon is the only material shown in Table 13.2 which has a negative temperature coefficient. The resistance of a piece of carbon will fall as it is heated.

Example 13.10 A carbon resistor has a resistance of 1000 Ω at 0°C. Find its resistance at 100°C. If a constant p.d. of 100 V is applied across the resistor, find the current flowing in the resistor at each temperature.

$$R = R_0(1 + \alpha\theta) = 1000(1 - 0.0005 \times 100)$$
$$= 1000(1 - 0.05)$$
$$= 1000 - 50$$
$$= 950\ \Omega$$

At 0°C $\qquad I = \dfrac{100}{1000} = 0.1\ \text{A}$

At 100°C $\qquad I = \dfrac{100}{950} = 0.1053\ \text{A}$

Question 13.10 A carbon resistor has a resistance of 20 kΩ at 20°C. Find (a) its resistance at 0°C and (b) its resistance at 80°C. If a constant p.d. of 500 V is applied across the resistor, find the power dissipated by the resistor (c) at 20°C and (d) at 80°C. (20 202 Ω; 19 596 Ω; 12·5 W; 12·78 W)

13.8 Semiconductor material

In Chapter 11, Section 11.12, a semiconductor was stated as having a resistance which decreased with rise of temperature; that is, a negative temperature coefficient.

Figure 11.16 showed that the temperature coefficient was not a constant but increased as the semiconductor was heated.

When a current flows in a semiconductor material heat will be caused ($W = I^2Rt$ joules). This heat, if not dissipated, will cause the resistance of the material to fall and the current to increase. The heat generated will increase and the resistance will fall even farther, and this process will continue until the heat produced is sufficiently great to destroy the structure of the material.

The thermistor

The thermistor is a semiconductor material used for regulating or measuring temperature. Some car thermometers operate on this principle, the gauge being an ammeter which is measuring the current in, and therefore the resistance of, the thermistor. The gauge can therefore be calibrated to indicate temperature.

Thermistors can also be buried at various positions within the winding of an electric motor. If the winding overheats at any point, the thermistor current will suddenly increase and operate a cut-out in the motor supply circuit. The thermistor can therefore be described as a non-linear resistor, which will *not* obey Ohm's law.

A device such as this can be used to counteract the effect of positive temperature coefficient, as described in Section 13.5.

Additional questions

13.11 A coil of wire has a resistance of 10 Ω at 0°C. If the temperature coefficient of resistance is 0·0043 Ω/Ω at 0°C/°C find the resistance at 10, 20, and 30°C.

13.12 A tungsten filament lamp has a resistance of 250 Ω at 0°C. Find (a) its resistance at 2000°C and (b) the current taken from the supply of 250 V at each temperature. Take the temperature coefficient of resistance as 0·005 Ω/Ω at 0°C/°C.

13.13 A cable 3 km long has two copper conductors, each of 2 mm diameter. Calculate the value of its resistance. If a d.c. supply, voltmeter, and ammeter were available, explain a practical method of measuring the resistance.

13.14 The resistance of a filament lamp is 270 Ω at 1685°C and 490 Ω at 0°C. Find the average temperature coefficient for its filament.

13.15 The resistance of a coil of copper wire measures 20 Ω at the normal atmospheric pressure boiling point of water. Find its resistance at the freezing point of water.

13.16 Explain the terms 'semiconductor', 'conductance', and 'non-linear resistance'. Calculate the resistance of 30 m of 30 s.w.g. (0·0315 cm diameter) manganin wire. Take the specific resistance of manganin as $4·07 \times 10^{-7}$ Ωm. (C.G.L.I.)

13.17 Explain what is meant by the terms 'resistivity' and 'conductivity'. The winding space of a bobbin is filled with 20 m of 0·093 cm diameter wire of copper with resistivity $1·60 \times 10^{-8}$ Ωm. What is the resistance of the coil? If the same winding space were filled with wire of half the diameter, what would be the new resistance? (C.G.L.I.)

13.18 (a) A coil made of copper wire, 0·25 mm diameter, is to have a resistance of 8·0 Ω. What length of wire must be used?
(b) If this coil were connected across a 6 V d.c. supply how much heat would be generated per hour? The resistivity of copper is $1·72 \times 10^{-8}$ Ωm. (C.G.L.I.)

13.19 Explain the term 'temperature coefficient of resistance'. The voltmeter in Fig. 13.6 has a resistance of 1000 Ω. When R is at a temperature of

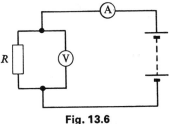

Fig. 13.6

0°C the meter readings are 1 V and 21 mA. At 40°C they are 1 V and 18·2 mA. What is the temperature coefficient of resistance of R?

(C.G.L.I.)

13.20 The resistance of a conductor 1 mm diameter and 100 m long is 2·2 Ω. What will be the resistance of a conductor made of the same material 0·5 mm diameter and 50 m long? (C.G.L.I.)

13.21 The resistance of a 12 V, 32 W filament lamp, as measured with an ohmmeter, is approximately 0·5 Ω. Compare this with the calculated value and explain the difference.

With the aid of a circuit diagram, describe the method you would use to measure the resistance of a 12 V lamp of unknown power rating, if the only available source had an e.m.f. of 20 V.

(C.G.L.I.)

13.22 The resistance of a copper coil is 15 Ω at 12°C. What is its resistance at 0°C given that the temperature coefficient of resistance referred to 0°C is 0·004 Ω/Ω at 0°C/°C? (C.G.L.I.)

13.23 A wire 10 m long and 0·05 cm diameter is made of copper having resistivity of $1·7 \times 10^{-8}$ Ωm at 20°C. The temperature coefficient of resistance is 0·004 Ω/Ω at 0°C/°C. Calculate the resistance of the wire at (a) 20°C, (b) 0°C, and (c) 50°C. (Based on C.G.L.I.)

13.24 For a laboratory experiment to verify Ohm's law a fixed resistor and a variable resistor are connected in series with a secondary cell. Also in the circuit are a voltmeter and an ammeter to measure the p.d. across the fixed resistor and the current through it.

(a) Sketch the circuit, including the voltmeter and ammeter, (b) explain fully what happens when the variable resistor is adjusted and the procedure by which Ohm's law may be verified, and (c) if during the experiment the fixed resistor rises in temperature due to the current passing through it, explain how this is likely to effect the results of the experiment. (C.G.L.I.)

14 The Use of Ammeters and Voltmeters

14.1 Ammeter–voltmeter range extension

In Chapter 12 it was stated that an ammeter has a low resistance (Section 12.3) and a voltmeter has a high resistance (Section 12.4). Most electrical measuring instruments in general use are basically ammeters. Their circuits are then arranged so that the instrument scale will indicate current, or potential difference, or resistance.

A basic ammeter is shown in Fig. 14.1.

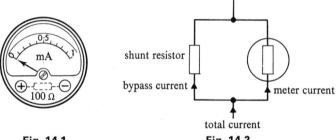

Fig. 14.1   Fig. 14.2

The ammeter indicator needle will read full-scale deflection when a current of 1 mA is flowing. The resistance of its movement is 100 Ω. With this instrument it is therefore possible to read up to 1 mA, but no higher. The p.d. across the instrument terminals will be

$$V = IR = 1 \times 10^{-3} \times 100 = 0{\cdot}1 \text{ V}$$

The instrument will thus read potential difference up to 0·1 volt.

Extending the current range
A resistor which is known as a 'shunt' is used to bypass the excess current. Figure 14.2 shows the general arrangement of the circuit.

Example 14.1 Calculate the value of shunt resistor required to enable the 1 mA full-scale deflection, 100 Ω basic instrument to read up to 10 mA.

From Fig. 14.2, as the circuit is a parallel arrangement, the p.d. across the meter and the shunt are equal. This p.d. is

$$V = IR = 1 \times 10^{-3} \times 100 = 0{\cdot}1 \text{ V}$$

The bypass current is

$$10 - 1 = 9 \text{ mA}$$

Therefore, the value of shunt resistor required must be

$$R = \frac{V}{I} = \frac{0 \cdot 1}{9 \times 10^{-3}} = \frac{100}{9} = 11 \cdot 11 \text{ }\Omega$$

Figure 14.3 shows the circuit and the calculated values.

Question 14.1 Find the value of shunt required to enable the 1 mA full-scale deflection, 100 Ω basic instrument to read up to 100 mA. (1·01 Ω)

The answers to Example 14.1 and Question 14.1 show that as the current increases the value of shunt resistance decreases. With very small values of shunt resistance it is more practical to wire the basic instrument across the shunt, as shown by Fig. 14.4, rather than as shown by Figs. 14.2 and 14.3.

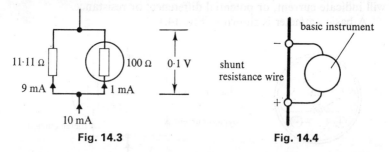

Fig. 14.3   Fig. 14.4

The resistance of the connecting wires to the instrument are negligible compared with the 100 Ω of the instrument, but this is not true with respect to the shunt resistance.

Extending the range of potential difference
A resistor which is known as a 'multiplier' is connected in series with the instrument so that the excess p.d. may be 'dropped' externally to the instrument. The general arrangement of the circuit is shown in Fig. 14.5.

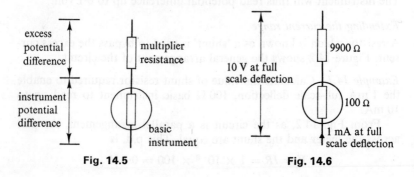

Fig. 14.5   Fig. 14.6

144 THE USE OF AMMETERS AND VOLTMETERS

Example 14.2 Calculate the value of multiplier resistance required to enable the 1 mA full-scale deflection 100 Ω basic instrument to read up to 10 V.

Figure 14.5 shows that the voltmeter circuit is a series circuit. The current throughout the circuit will thus be the same. At the full-scale deflection of the instrument—that is, when the reading is to be 10 V—the current in the circuit will be 1 mA.

The total resistance of the circuit is

$$R = \frac{V}{I} = \frac{10}{1 \times 10^{-3}} = 10 \text{ k}\Omega$$

Therefore, the multiplier resistance must be:

$$\text{Total resistance} - \text{Meter resistance}$$

that is,

$$10\,000 - 100 = 9900 \ \Omega$$

Figure 14.6 shows the circuit and the calculated values.

Question 14.2 Calculate the value of multiplier resistance required to enable the 1 mA full-scale deflection, 100 Ω basic instrument to read 100 V at full-scale deflection. (99 900 Ω)

14.2 The effect on a circuit of the application of a voltmeter

Figure 14.7 shows a circuit having two 10 Ω resistors connected in series across a 2 V supply. The current flowing in the circuit will be

$$I = \frac{V}{R} = \frac{2}{10 + 10} = 0\cdot 1 \text{ A}$$

The p.d. of each resistor will be

$$V = IR = 0\cdot 1 \times 10 = 1 \text{ V}$$

If the voltmeter of Example 14.2 is used to measure the p.d. of resistor R_1, as shown by Fig. 14.8, the circuit will be changed.

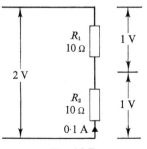

Fig. 14.7

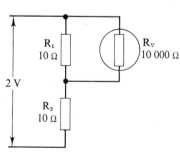

Fig. 14.8

The effective resistance of the voltmeter R_v and R_1 is

$$R = \frac{R_1 \times R_v}{R_1 + R_v} = \frac{10 \times 10\,000}{10 + 10\,000} = \frac{100\,000}{10\,010}$$

$= 10\ \Omega$ by the use of four figure reciprocal tables.

The total circuit resistance is

$$Rt = R + R_2 = 10 + 10 = 20\ \Omega$$

therefore,

$$I = \frac{V}{Rt} = \frac{2}{20} = 0.1\ \text{A}$$

As the measured value of the resistor is low compared with the voltmeter resistance, the effect on the circuit is negligible.

Example 14.3 (i) Two resistors, each of 10 000 Ω resistance, are connected in series to a 2 V supply. Find (a) the circuit current, (b) the p.d. across each resistor, and (c) the power dissipated by each resistor.

(ii) A voltmeter having a resistance of 10 000 Ω is then connected across one of the resistors. Find how the values of (a), (b), and (c) of part (i) are affected.

Figure 14.9 shows the circuit arrangement without the voltmeter.

(i) (a) The total resistance of the circuit is

$$R_1 + R_2 = 10 + 10 = 20\ \text{k}\Omega$$

therefore,

$$I = \frac{V}{R} = \frac{2}{20 \times 10^{-3}} = 0.1\ \text{mA}$$

(b) Potential difference of each resistor is

$$I \times R = 0.1 \times 10^{-3} \times 10 \times 10^3 = 1\ \text{V}$$

(c) Power dissipated by each resistor is

$$I \times V = 0.1 \times 10^{-3} \times 1 = 0.1\ \text{mW}$$

Figure 14.10 shows the circuit arrangement with the voltmeter connected across R_1.

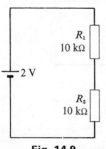

Fig. 14.9

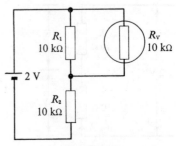

Fig. 14.10

(ii) (a) The total resistance of R_1 and R_v in parallel is

$$R = \frac{R_1 \times R_v}{R_1 + R_v} = \frac{10 \times 10}{10 + 10} = \frac{100}{20} = 5 \text{ k}\Omega$$

The resistance of whole circuit is

$$Rt = R_2 + 5 \text{ k}\Omega = 10 + 5 = 15 \text{ k}\Omega$$

$$I = \frac{V}{R} = \frac{2}{15} \times 10^{-3} = 0.133 \text{ mA}$$

Therefore the current in R_2 will be

$$0.133 \text{ mA}$$

(b) The p.d. of R_2 is

$$I \times R = 0.133 \times 10^{-3} \times 10 \times 10^3 = 1.33 \text{ V}$$

The p.d. of R_1 and R_v is

$$I \times R = 0.133 \times 10^{-3} \times 5 \times 10^3 = 0.666 \text{ V}$$

The current in R_1 is

$$\frac{V}{R} = \frac{0.666}{10 \times 10^3} = 0.0666 \text{ mA}$$

(c) The power dissipated by each resistor is IV, thus the power dissipated by R_1 is

$$0.0666 \times 10^{-3} \times 0.666 = 0.0443 \text{ mW}$$

and the power dissipated by R_2 is

$$0.133 \times 10^{-3} \times 1.33 = 0.177 \text{ mW}$$

These values can be shown more clearly on a circuit diagram (Fig. 14.11), and a comparison of the answers to part (i) and part (ii) of Example 14.3 is given in Table 14.1.

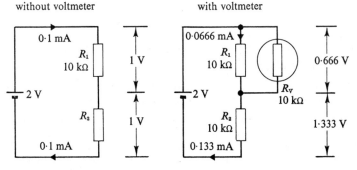

Fig. 14.11

THE EFFECT ON A CIRCUIT

From Table 14.1 and Fig. 14.11 it will be seen that the inclusion of the voltmeter has completely changed the characteristics of the circuit. As the voltmeter readings are nowhere near the true p.d. of the resistors, the test is valueless. The power rating of a resistor could, in an exceptional case, be exceeded.

Thus not only is it a waste of time to use a voltmeter on some circuits but this may also damage the circuit. It is therefore important to observe the rule:

Never connect a voltmeter across a resistance which is high compared with the resistance of the voltmeter.

Question 14.3 Two resistors, each of resistance 100 kΩ, are connected in series to a battery of e.m.f. 10 V. The p.d. of one of the resistors is measured with the 10 V, 10 kΩ voltmeter of Example 14.2. Calculate (a) the p.d. across each resistor, (b) the current in each resistor, and (c) the power dissipated by each resistor, (i) before and (ii) after the connection of the voltmeter.

((i) (a) 5 V; (b) 50 μA; (c) 250 μW)
((ii) (a) 9·164 V, 0·836 V; (b) 91·64 μA, 8·33 μA; (c) 6·963 μW, 839·9 μW)

The expression 'ohms per volt'

The basic ammeter considered in the examples so far has had a full-scale deflection of 1 mA and a resistance of 100 Ω. The p.d. across its coil at full-scale deflection is $V = IR = 1 \times 10^{-3} \times 100 = 0\cdot1$ V. The ratio of resistance of coil to p.d. required for full-scale deflection, which in this case is $100/0\cdot1 = 1000$, is termed the 'ohms per volt' of the instrument.

This value is used to determine quickly the resistance of the voltmeter with any value of multiplier connected.

Example 14.4 Find the resistance of a 1 mA full-scale deflection 100 Ω instrument when it is modified to read 10 V full-scale deflection.

The p.d. of the instrument is

$$IR = 1 \times 10^{-3} \times 100 = 0\cdot1 \text{ V}$$

The ohms per volt is

$$\frac{R}{V} = \frac{100}{0\cdot1} = 1000 \ \Omega/V$$

When reading 10 V full-scale deflection the meter resistance will be $1000 \times 10 = 10$ kΩ. This compares with the calculation of total resistance in Example 14.2.

Question 14.4 Find, for the basic instrument of Example 14.4, the resistance of the meter when converted to read 100 V full-scale deflection. (100 kΩ)

Question 14.5 An ammeter has a full-scale deflection of 5 mA and a resistance of 50 Ω. The instrument is then converted to a voltmeter having a full-scale deflection of 500 V. Find (a) the value of multiplier required, (b) the value of its ohms per volt rating, and (c) the total resistance of the voltmeter.
(99 950 Ω, 200 Ω/V, 100 kΩ)

Table 14.1

Items	Without voltmeter	With voltmeter
Potential difference R_1	1 V	0·666 V
Potential difference R_2	1 V	1·33 V
Current in R_1	0·1 mA	0·0666 mA
Current in R_2	0·1 mA	0·133 mA
Power dissipated by R_1	0·1 mW	0·0443 mW
Power dissipated by R_2	0·1 mW	0·177 mW

14.3 A simple ohmmeter

Figure 14.12 shows the method of connection of an ammeter for the measurement of resistance. If the battery used has an e.m.f. of 1·5 V,

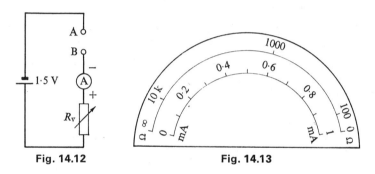

Fig. 14.12   **Fig. 14.13**

and the meter has a full-scale deflection of 1 mA and resistance 100 Ω, then when the terminals A and B are short circuited and R_v is adjusted so that the instrument reads full-scale deflection, the total resistance of the circuit will be:

$$R = \frac{V}{I} = \frac{1 \cdot 5}{1 \times 10^{-3}} = 1 \cdot 5 \text{ k}\Omega$$

The value of R_v will need to be 1500 − 100 = 1400 Ω.

When full-scale deflection is registered the measured resistance between terminals A and B must be zero. If a resistor is connected between terminals A and B the current in the meter will be reduced from the full-scale deflection current. When the meter reading is zero the terminals A and B must be open circuited, the external resistance being infinity.

Example 14.5 Calculate the current flowing in the ohmmeter of Fig. 14.12 when resistors of 100 Ω, 1000 Ω, and 10 000 Ω are connected between terminals A and B.

100 Ω: $I = \dfrac{V}{R} = \dfrac{1 \cdot 5}{1500 + 100} = \dfrac{1 \cdot 5}{1600} = 0 \cdot 938$ mA

1000 Ω: $I = \dfrac{V}{R} = \dfrac{1 \cdot 5}{1500 + 1000} = \dfrac{1 \cdot 5}{2500} = 0 \cdot 6$ mA

10 000 Ω: $I = \dfrac{V}{R} = \dfrac{1 \cdot 5}{1500 + 10\,000} = \dfrac{1 \cdot 5}{11\,500} = 0 \cdot 13$ mA

The ohmmeter scale will now appear as shown in Fig. 14.13.

Question 14.6 Add to the scale of resistance of Fig. 14.13 the resistance values when the current flowing in the ammeter is (a) 0·8 mA, (b) 0·5 mA, and (c) 0·2 mA. (375 Ω, 1·5 kΩ, 6 kΩ)

Accuracy of the simple ohmmeter

The accuracy of the ohmmeter will depend upon the accuracy of the battery e.m.f. As in service this will vary with age; this type of ohmmeter can only be used for estimating resistance values.

Example 14.6 Calculate new current values for Example 14.5 if the battery e.m.f. has fallen to 1·3 V.

The ohmmeter would be adjusted to full-scale deflection with terminals A and B short circuited.

When R external = 0,

$$\text{Resistance of circuit} = \dfrac{V}{I} = \dfrac{1 \cdot 3}{1 \times 10^{-3}} = 1 \cdot 3 \text{ k}\Omega$$

When R external = 100 Ω,

$$I = \dfrac{V}{R} = \dfrac{1 \cdot 3}{1300 + 100} = \dfrac{1 \cdot 3}{1400} = 0 \cdot 929 \text{ mA}$$

When R external = 1000 Ω,

$$I = \dfrac{V}{R} = \dfrac{1 \cdot 3}{1300 + 1000} = \dfrac{1 \cdot 3}{2300} = 0 \cdot 565 \text{ mA}$$

When R external = 10 000 Ω,

$$I = \dfrac{V}{R} = \dfrac{1 \cdot 3}{1300 + 10\,000} = \dfrac{1 \cdot 3}{11\,300} = 0 \cdot 115 \text{ mA}$$

Table 14.2 shows a comparison of the answers of Examples 14.5 with 14.6.

Question 14.7 Find the resistance values for Question 14.6 when the battery potential is 1·3 V. (325 Ω, 1·3 kΩ, 5·2 kΩ)

14.4 The multi-range meter

A multi-range meter consists of a basic milli-ammeter or micro-ammeter with various shunt and multiplier resistors that may be switched into

Table 14.2

Resistance, Ω	Current when $V = 1\cdot5$ V, mA	Current when $V = 1\cdot3$ V, mA
0	1	1
100	0·947	0·929
1000	0·6	0·565
10 000	0·13	0·115
∞	0	0

the circuit as desired. It is also customary to include a small dry battery to enable the instrument to be used to estimate resistance values. If the instrument is to read both alternating current and direct current, a rectifier would be fitted into the circuit.

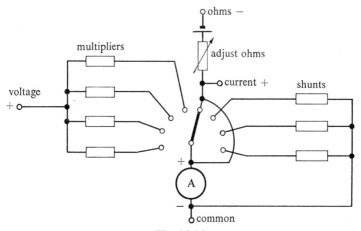

Fig. 14.14

The circuit diagram (Fig. 14.14) shows how the shunts, multipliers, and resistance measuring circuits could be brought together by the addition of a switch to form a d.c. multi-range meter.

Additional questions

14.8 A multi-range ammeter is fitted with three shunts to give the following ranges: 0–2·5 mA, 0–10 mA, 0–50 mA, and 0–250 mA. If the resistance of the ammeter is 75 Ω, find the value of the required shunt resistors. With the aid of a circuit diagram show how they could be connected.

14.9 A multi-range voltmeter has five ranges: 0–2·5 V, 0–10 V, 0–25 V, 0–100 V, and 0–250 V. If the basic instrument has a full-scale deflection of 500 μA and a resistance of 150 Ω, find the value of multiplier resistors required. Draw a circuit diagram to show the method of connection.

14.10 A simple ohmmeter is constructed using an ammeter with a full-scale deflection of 5 mA, a variable resistor, and a 3 V battery. Draw a suitable circuit for the arrangement.

Calculate the resistance indicated by the instrument when currents of (a) 0, (b) 2 mA, (c) 4 mA, and (d) 5 mA are flowing in the circuit.

14.11 An ammeter and a voltmeter are used to check the resistance of a resistor which is marked 1 MΩ. The circuit is connected as in Fig. 14.15. If the voltmeter has a resistance of 1 MΩ, find the resistance value from the calculated instrument readings. The resistance of the ammeter may be ignored. State the error in the calculated value.

14.12 A 20 Ω resistor is connected in series with an ammeter across a 24 V d.c. supply. The ammeter consists of a coil of 12 Ω resistance with a shunt resistor of 1·0 Ω. Across the 20 Ω resistor is connected a voltmeter consisting of a coil of 12 Ω resistance with a series resistor of 100 Ω. (i) Draw the circuit and (ii) calculate the current through each of the coils and the three resistors.   (C.G.L.I.)

14.13 A moving coil instrument gives full-scale deflection when the current through it is 15 mA and the p.d. across it is 6 V.
(a) Calculate the resistance of the instrument.
(b) Find the value of resistor required to enable the instrument to read 24 V at full-scale deflection and show by a sketch how it should be connected.
(c) Find the value of resistor required to enable the instrument to read 1·0 A at full-scale deflection and show by a sketch how it should be connected.   (C.G.L.I.)

14.14 A multi-range meter has 2 V, 6 V, 20 V, and 60 V ranges and is labelled '2000 ohms per volt'. The resistance of the movement is 40 Ω. Describe how these ranges are achieved. What would you expect the lowest current range of the meter to be?

A voltmeter used to measure the p.d. between two points gives a higher reading when switched to the 60 V range than when switched to the 2 V range. Explain this.   (C.G.L.I.)

14.15 Show how a meter having a full-scale deflection of 1 mA and a resistance of 50 Ω may be adapted for use as (a) a meter with full-scale deflection of 1 A and (b) a voltmeter with full-scale deflection of 1 V.

A simple ohmmeter is constructed by connecting the 1 mA meter in series with a 1·5 V cell and a resistor R, as shown in Fig. 14.16.

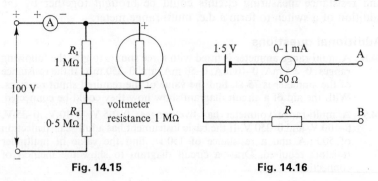

Fig. 14.15   Fig. 14.16

R is chosen so that full-scale deflection is obtained when the terminals A and B are shorted. Find the value of R. What is the value of resistance which, when connected between A and B, causes a reading of 0·65 mA?
(C.G.L.I.)

14.16 A voltmeter with resistance of 20 kΩ is used to measure the p.d. between points P and Q of the circuit shown in Fig. 14.17. What is the voltmeter reading, and how does this compare with the true value?

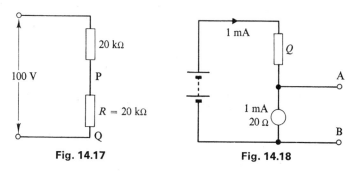

Fig. 14.17   Fig. 14.18

In the same circuit, the ammeter reading is taken while the voltmeter is still connected, and the two meter readings are used to calculate the value of R. What is the ammeter reading and what is the percentage error in the value of R? (C.G.L.I.)

14.17 The resistance of a 1 mA meter is 20 Ω. Show how the meter may be adapted to measure currents up to 10 mA.

Figure 14.18 shows one way in which the 1 mA meter may be used to measure resistance. The resistance Q reduces the current from the battery to 1 mA and the unknown resistance, connected between A and B, has negligible effect on this current. What resistance values connected between A and B will reduce the meter current to (a) ¼ mA, (b) ½ mA, and (c) ¾ mA? Sketch the meter scale calibrated in resistance values. What is the value of Q if the terminal voltage of the battery is 20 V? (C.G.L.I.)

14.18 An ammeter of resistance 0·1 Ω has a shunt of resistance 0·03 Ω connected across it. What is the total current flowing when the current in the ammeter is 2·1 A? (C.G.L.I.)

15 The Chemical Effect of an Electric Current

15.1 Electrolysis

When an electric current passes through some substances, chemical changes occur. Such a material is called an **electrolytic conductor.**

Electrolyte

One type of electrolytic conduction is termed **ionic** and a substance with this property is called an **electrolyte.** Two well-known electrolytes are copper sulphate and sulphuric acid.

Ions

In electrolyte solutions the atoms are very loosely bonded electrostatically, and freely interchange. Some of the atoms, as they wander about, take with them additional electrons, and these atoms are said to be negatively charged **ions.** The remainder of these electrolyte atoms are therefore short of electrons and are called positive ions.

Electrolytic cell

Figure 15.1 shows an electrolytic cell. Two carbon electrodes are inserted into a vessel containing sulphuric acid and distilled water. When a battery is connected across the cell terminals, one of the electrodes will have a positive charge and the other a negative charge.

The external d.c. supply will pass current through the cell from the positive to the negative electrode in a conventional sense.

Cathode and anode

The two electrodes are termed **cathode** and **anode.** The cathode is defined as being the conductor by which *the current (conventional) leaves the electrolyte.* It follows, therefore, that the anode must be the other electrode.

The electrolysis of water

When the current passes from the anode to the cathode of the electrolytic cell shown in Fig. 15.2, the positive ions of the electrolyte, which are hydrogen atoms, will be attracted to the negative electrode or cathode. Here the positive charge will be lost and the hydrogen will leave the electrolyte in the form of gas.

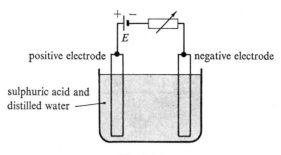

Fig. 15.1

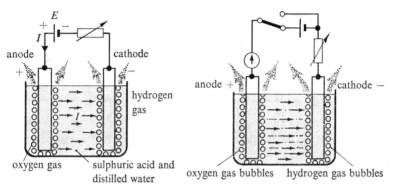

Fig. 15.2   **Fig. 15.3**

In a similar way, when the negative ion—which is attracted to the positive electrode or anode—discharges, the gas given off will be oxygen. As pure water is formed from hydrogen and oxygen, the water will gradually be lost from the cell.

If the circuit of Fig. 15.3 is now connected to the electrolytic cell the following will occur: When the battery is passing current through the cell the galvanometer will indicate in one direction, but when the switch is changed over to disconnect the battery, the galvanometer, which is still connected across the cell terminals, will have a reversed reading. This reverse current will continue until the gas bubbles on the electrodes revert into the electrolyte. The process is thus reversible.

15.2 Faraday's laws of electrolysis

Michael Faraday discovered in 1833 that the quantity of the substance liberated depended on the number of electrons gained by the negative ion and lost by the positive ion.

This law was later presented in a more practical way: *The mass of a substance liberated from an electrolyte depends upon (1) the quantity of electricity passed, Q coulombs, and (2) the type of substance.*

Electrochemical equivalent (E.C.E.)

The two conditions of the laws are normally combined in a constant called the electrochemical equivalent, of which a typical list is given in Table 15.1.

Table 15.1

Material	Electrochemical equivalent, mg/C	Material	Electrochemical equivalent, mg/C
Aluminium	0·0936	Lead	1·072
Carbon	0·0623	Nickel	0·304
Chromium	0·0898	Oxygen	0·0829
Copper	0·330	Silver	1·118
Gold	0·6802	Tin	0·615
Hydrogen	0·010 44	Zinc	0·339
Iron	0·193		

Example 15.1 Find the mass of hydrogen released from the electrolytic cell shown in Fig. 15.2 when a constant current of 3 A is passing through the cell for a time of 2 h.

Quantity of electricity passed in 2 h will be

$$Q = I \times t$$
$$= 3 \times 2 \times 60 \times 60 = 21\,600 \text{ C}$$

From Table 15.1, 1 coulomb of electricity will release 0·010 44 mg of hydrogen. Therefore, the mass released by 21 600 coulombs will be:

$$0·010\,44 \times 21\,600 = 225·5 \text{ mg}$$

This calculation may be made by using the formula:

$$m = zIt$$

where z is the electrochemical equivalent. Thus,

$$m = 0·010\,44 \times 3 \times 2 \times 60 \times 60$$
$$= 225·5 \text{ mg}$$

Question 15.1 Find the mass of oxygen released by the electrolytic cell shown in Fig. 15.1 by a constant current of 5 A flowing for 6 h.   (8·95 g)

15.3 Practical use of electrolytic action
Electric cell

There are two basic types of electric cell:

(1) *The primary cell* is an electric cell which is made by the combination of specific chemicals. When the cell is used to supply electrical

energy, chemical changes will occur and when the action eventually ceases, the cell is discarded.

The Leclanché cell is the most commonly used primary cell. The construction of a Leclanché cell can be seen in Fig. 15.4 (wet) and Fig. 15.5 (dry), while Fig. 15.6 illustrates the cell's chemical action.

The need for a depolarizer From Fig. 15.6 it can be seen that without the depolarizer, hydrogen gas would be given off at the positive electrode which in this case, by its definition, is the cathode. The gas would collect around the cathode in the form of bubbles and would thus separate the electrode from the electrolyte, breaking the electrical circuit.

The purpose of a depolarizer is to change hydrogen into water; thus the manganese dioxide depolarizer must have a high oxygen content. In practice, the depolarizer is the largest part of the cell, and the maximum current that the cell will provide will depend on the efficiency of this unit.

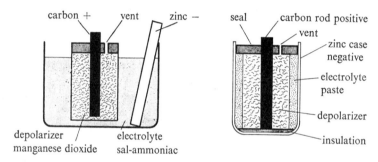

Fig. 15.4 Wet Leclanché cell   Fig. 15.5 Dry Leclanché cell

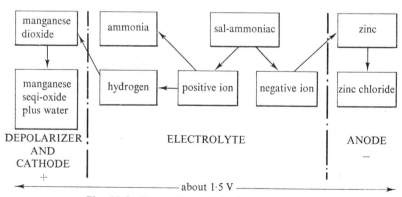

Fig. 15.6 The action of the Leclanché cell

In the form shown in Fig. 15.4 the Leclanché cell is not portable. In the so-called dry cell (Fig. 15.5) the liquid is suspended in paste form. The depolarizer is a mixture of manganese dioxide and crushed carbon—the latter being added to reduce the resistance of the cell—and the electrolyte is a paste made from sal-ammoniac and zinc chloride mixed with flour and gum.

Use of the Leclanché cell Owing to the effect of polarization, the Leclanché cell, wet or dry, can only be used in circuits requiring a small continuous current, or in circuits requiring short bursts of larger currents.

Example 15.2 State two uses of the Leclanché cell, one continuous the other intermittent.

    Continuous: Transistor radio battery.
    Intermittent: Gas lighter battery.

Question 15.2 State two further uses for the Leclanché cell in which one is continuous and the other intermittent.

Local action Owing to the presence of traces of iron in the commercial form of zinc, local short-circuited cells are caused. The zinc is thus eaten away without the cell being in use. In the case of the wet type of Leclanché cell it is possible to lift out the zinc electrode. Alternatively, the zinc may be cleaned and dipped in mercury to seal the surface. This type of zinc is called amalgamated zinc.

The Weston reference cell
The **Weston** cell, which is a primary cell, is used as an e.m.f. reference. The e.m.f. without any current drain will be 1·018 48 V at 15°C and 1·018 30 V at 20°C.

    Mercury is used for the positive electrode and cadmium for the negative electrode. The electrolyte is cadmium sulphate and the depolarizer is mercurous sulphate.

(2) *The secondary cell*
A **secondary** cell has to be **charged**—that is, an electric current is passed into the cell, changing the chemicals in the process, and storing the electrical energy. When required the electrical energy can be released, the process being termed **discharge**. These processes are shown in Fig. 15.7.

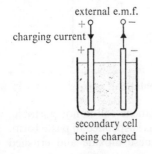
secondary cell being charged

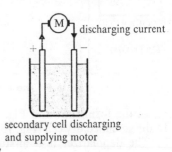
secondary cell discharging and supplying motor

Fig. 15.7

The lead–acid cell

The lead–acid cell is very widely used becaue it is efficient and relatively cheap. Figure 15.8 shows the chemical arrangement and how the chemicals change.

Following the diagram downwards shows the method by which the cell changes as it gives up its stored electrical energy. If the diagram is followed upwards, this will show the chemical changes as the cell is charged. From this diagram it will be seen that the positive plate has a greater number of chemical reactions than the negative plate. Figure 15.9 shows how the plates are arranged in the casing. The plates are interleaved so that each positive plate will have a negative plate on either side.

Each side of the positive plate will be exposed to the chemical reactions and the heat produced will expand the plate evenly and thus prevent buckling.

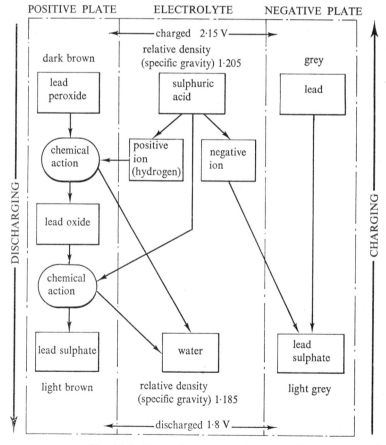

Fig. 15.8 The action of a lead–acid cell

PRACTICAL USE OF ELECTROLYTIC ACTION

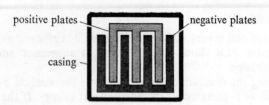

Fig. 15.9 Plan view of lead–acid cell showing arrangement of plates

The plates can be produced by two methods:

(1) *The Planté or formed method.* The plates are made from lead and usually they would be finned rather like the fins of an air-cooled motor cycle engine to increase their surface area within the electrolyte. The cell is charged and discharged continuously until the lead takes on the chemical form shown in Fig. 15.8.

(2) *Pasted plates* (Faure process). A cheaper and quicker method of construction is to force the prepared chemicals into a lead framework.

The formed method produces a very strong and reliable cell, but owing to its high cost it is only used for very special applications. As a compromise, cells are sometimes constructed with a formed positive and a pasted negative. The plates are separated by glass, celluloid, or wood to prevent short-circuiting by buckling or mechanical vibration. A space is left below the plates to permit sludge to accumulate without short-circuiting the plates. The case is made of glass or tough rubber, or for some open-top cells, of lead-lined wooden boxes.

Use of the lead–acid cell   Lead–acid cells are rated according to the quantity of electricity they can store, and the time of discharge.

Example 15.3   A lead–acid cell is rated at 100 ampere-hours (Ah) at a 20 h rate. Find the maximum discharge current and the quantity of electricity stored by the cell in coulombs.

$$\text{Quantity} = \text{Current} \times \text{Time}$$

If the current is in amperes and the time is in hours then the quantity will be in ampere-hours:

$$\text{Current} = \frac{\text{Quantity}}{\text{Time}} = \frac{100}{20} = 5 \text{ A}$$

Quantity in coulombs = I ampere × t seconds

$$= 5 \times 20 \times 60 \times 60$$
$$= 360\,000 \text{ C}$$

Question 15.3   A 44 Ah battery is rated at a 10 h rate. Find (a) the maximum discharge current and (b) the total energy stored, in joules, if the p.d. of the battery terminals remains constant at 12 V.   (4·4 A, 1·9 × 10⁶ J)

The lead–acid secondary cell has a very low internal resistance, and extremely high intermittent currents may therefore be taken from the cell.

Advantage is taken of this fact in its use for starting automobile engines. The current required will vary considerably but may be in excess of 200 A.

Other uses for lead–acid cells are found in the telecommunications field where they are used extensively for telephone and telegraph circuits.

Testing for state of charge Figure 15.8 shows that the state of charge can be checked:

(1) by the relative density (specific gravity) of the electrolyte,
(2) by the colour of the plates, and
(3) by the p.d. of each cell.

Care of lead–acid cells The following rules must be observed if cells are to have a reasonable length of life.

(1) Cells should not be left in a discharged state as the plates become covered with a substance known as white sulphate which will effectively reduce their surface area.
(2) A periodic overcharge will help to remove white sulphate.
(3) The cell should be topped up to a level above the plates with distilled water.
(4) The manufacturer's charging procedure should be observed.
(5) The temperature of the cell should not be allowed to exceed 40°C.

The alkaline type of secondary cell

The alkaline type of secondary cell is used extensively for traction duties —for example, driving electric delivery vans and fork lift trucks—for which purposes it is better suited than the lead–acid cell.

There are two types of alkaline battery in common use: the nickel–cadmium cell and the nickel–iron cell. Since both these cells are very similar, only the nickel–iron cell will be considered here.

The nickel–iron alkaline cell

The positive plate of the nickel–iron alkaline cell is formed from alternate layers of nickel hydroxide and either flake nickel or graphite, enclosed in perforated nickel-plated mild steel tubes. The purpose of the flake nickel or graphite is to reduce the resistance as nickel hydroxide is a poor electrical conductor.

The negative plate is formed from iron oxide and a small amount of mercuric oxide, again to reduce resistance, enclosed in perforated steel pockets which are assembled in nickel-plated steel plates.

The positive and negative plates are contained in a nickel-plated steel container which will also contain the electrolyte, potassium hydroxide. As the steel case is 'live' each cell must also be enclosed in an insulating crate.

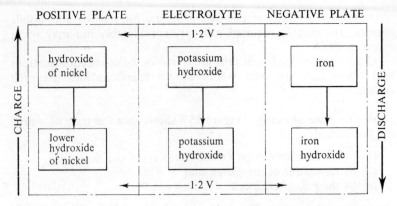

Fig. 15.10 Action of a nickel–iron alkaline cell

The action of the cell is shown in Fig. 15.10 and it will be seen from this diagram that the action is less involved than with the lead–acid cell. The figure shows that the electrolyte is unchanged chemically between charged and discharged conditions, and therefore its relative density is constant. The p.d. across the plates is also constant, about 1·2 V, and thus the relative density and p.d. will not give any indication of the level of charge.

The nickel–iron cell is light, mechanically strong, and free from corrosive fumes. The charge and discharge rates can be high and the cell may be stored without deterioration. In comparison with the lead–acid cell it has a higher capital cost, a lower e.m.f., and a lower capacity.

Question 15.4 Compare the advantages and disadvantages of the lead–acid and nickel–iron alkaline cells, and give two possible uses of each.

Question 15.5 Explain the difference between a primary and a secondary cell. Say which of these two types of electric cell are normally used in (a) a car, (b) a milk float, (c) an electric torch, and (d) a transistor radio set.

15.4 Electroplating

Figure 15.11 shows, diagrammatically, an electroplating bath arranged for copper plating.

A bar of copper and the metal article to be plated are placed in a bath containing a solution of copper sulphate. A direct current supply is joined across the two metals, the article to the negative and the copper to the positive of the supply.

When an electric current is passed through the plating bath, the copper sulphate will split into positive (copper) ions and negative ions. The copper will be attracted to the negative electrode, which is the article to be plated, and will lose its charge, the copper atom remaining on the article. The negative ion will be attracted to the positive electrode (which is the copper bar), and as it discharges, an atom of copper will be removed to reform as copper sulphate. As the current leaves the

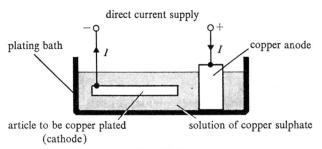

Fig. 15.11

electrolyte at the plated article, this will be the cathode—the copper being the anode. The cathode will increase in mass and the anode will lose an equal amount of mass.

Example 15.4 A piece of steel is placed in a copper-plating bath and a current of 1 A is passed for 2 h. Find the mass of copper deposited on the steel.

From Table 15.1 the electrochemical equivalent of copper is 0·33 mg/C.

$$m = zIt = 0·33 \times 1 \times 2 \times 60 \times 60 = 2375 \text{ mg}$$
$$= 2·375 \text{ g}$$

Question 15.6 How long must a piece of steel be left in a copper-plating bath if a constant current of 2 A is passing through the bath and the amount of deposit required is 10·5 g? (4 h 26 min)

Example 15.5 A piece of steel, which has been copper plated, is to have a further 125 μm (125 micron) layer of nickel. If the steel has a total surface area of 80 cm² and the current passing through the nickel-plating bath is to be 4 A, find the period of time for which the current must be maintained.

Volume of nickel plating = Surface area × Thickness of plating

$$= 80 \times \frac{125}{10\,000} = 1 \text{ cm}^3$$

From Table 1.3 (Chapter 1) the relative density of nickel is 8·8.

$$\text{Relative density of nickel} = \frac{\text{Mass of nickel}}{\text{Mass of equal volume of water}}$$

Therefore

Mass of nickel = Relative density × Mass of equal volume of water
$$= 8·8 \times 1 = 8·8 \text{ g}$$

(1 cm³ of water has a mass of 1 g)

$$m = zIt$$

ELECTROPLATING

From Table 15.1, the electrochemical equivalent (z) for nickel is 0·304 mg/C, therefore

$$t = \frac{m}{zI} = \frac{8\cdot 8 \times 1000}{0\cdot 304 \times 4} = 7230 \text{ s} = \frac{7230}{60} = 120 \text{ min } 30 \text{ s}$$
$$= \text{very nearly 2 h}$$

Question 15.7 The piece of steel in Example 15.5 is now to be chromium plated to a thickness of 100 μm (100 micron). If the current is maintained at 3 A find the time taken. (5 h 45 min)

Many everyday articles have a plating finish. Car fittings that are made of steel are often plated with copper, then nickel, and finally chromium. Steel cans for food are plated with tin, and tableware with nickel then silver.

Before plating the article must be clean and all oxide must be removed.

15.5 Electrolytic corrosion

Figure 15.12 is a sketch map showing an electrified railway supplied by direct current. Crossing the railway at point A is a petroleum pipeline, and crossing the pipeline at point B, and also passing under the railway track at point C, is a lead sheathed telephone cable.

The steel railway track may have many high resistance joints and some current will leave the track to flow in the earth to the lead cable at point C. At point B the current passes into the pipeline and at point A it returns to the railway.

This circuit is in fact a parallel circuit with two branches. As the earth acts as an electrolyte the lead cable will be an anode at B, and will therefore lose metal. The pipeline will similarly lose metal at point A. Eventually the metals will be corroded to such an extent that a hole will develop in the sheathing.

In order to prevent the current from flowing in the cable, an 'insulating gap' is placed in the sheathing between B and C. An insulating gap is an insulating section, inserted into the cable, which will greatly increase the resistance of this branch of the parallel circuit and force the current to find some alternative path.

Example 15.6 Find the iron lost from the pipe at point A (Fig. 15.12), in a time of 90 days if the earth current flowing is 20 mA.

$$\text{Time} = 90 \times 24 \times 60 \times 60 \text{ s}$$
$$m = zIt = 0\cdot 193 \times \frac{20}{1000} \times 90 \times 24 \times 60 \times 60 \text{ mg}$$
$$= 30\,000 \text{ mg}$$
$$= 30 \text{ g}$$

Question 15.8 An earth current flowing in a lead sheathed cable is measured as 20 mA. Find the mass of lead lost in a period of 90 days at the position where the current leaves the cable. (167 g)

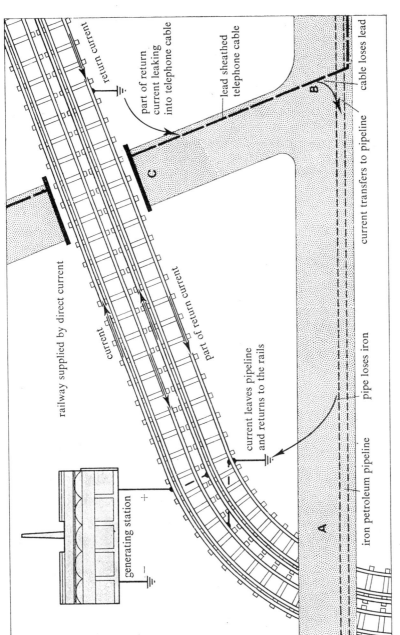

Fig. 15.12

Additional questions

15.9 An electrolytic cell consists of two carbon rods immersed in an electrolyte of dilute sulphuric acid. If a current of 2 A is passed through the cell for a time of 5 h, find (a) the mass of oxygen released and (b) the mass of hydrogen released.

15.10 A steel car bumper has a total surface area of 4900 cm² and is to have a layer of 100 μm of copper plated over its surface. The plating current is to be 15 A. Find the time that the plating process must be continued.

15.11 The copper-plated steel bumper of Question 15.10 is next to have a 100 μm layer of nickel plate. The plating is scheduled to take a time of $1\frac{1}{2}$ h. Find the current required.

15.12 The car bumper of Question 15.11 is finally to have a 125 μm layer of chromium plate. Find the time required for the process if the current through the plating bath is maintained at a constant value of 150 A.

15.13 A current of 1·5 A is passed through a solution of copper sulphate in an electrolytic cell. How long will it take to deposit 2·0 g of copper on the cathode, given that the electrochemical equivalent of copper is 0·33 mg/C? (C.G.L.I.)

15.14 What is meant by 'electrochemical equivalent'? Write short notes on two chemical effects of a current which are of practical importance. (C.G.L.I.)

15.15 State the type of battery you would use for each of the following applications, giving reasons for your choice. Calculate in each case the current supplied by the battery.
  (a) A 10 μF capacitor is charged to a p.d. of 48 V, and then fully discharged, five times a second.
  (b) A 48 V electric motor is used to drive a machine with an output of $\frac{1}{8}$ hp, the overall efficiency being 60 per cent. (1 hp = 0·746 kW.) (C.G.L.I.)

15.16 Make a sketch of a primary cell, showing the construction. Name the type of cell and indicate the components. (C.G.L.I.)

15.17 How long will it take to deposit 25 g of silver in an electroplating process using a current of 1·5 A? The electrochemical equivalent of silver is 1·12 mg/C.

15.18 When a current of 3 A is passed for 20 min through an electrolytic cell, 4·03 g of metal is deposited on the cathode. Find the amount deposited when a current of 1 A is passed for 30 min, and calculate the electrochemical equivalent of the metal.

  What battery would you use to supply the current, if the resistance of the electrolytic cell is 3 Ω? Give a summary of the properties of this battery. (C.G.L.I.)

15.19 When a current of 1 A is passed through an electrolytic cell for 1 h, 8·06 g of metal is deposited at one electrode and 1·99 g of gas is liberated at the other. What is the electrochemical equivalent of the gas?

  How long would it take to deposit 0·77 g of the metal using a current of 0·25 A, and what mass of the gas would be liberated in the process?

  Explain how electrolysis plays a part in corrosion. (C.G.L.I.)

16 Magnetism

Magnetism is not a new science. The Ancient Greeks knew the properties of the magnetic oxide of iron called lodestone. In the Middle Ages, magnetized iron was used for the navigators compass; later, William Gilbert, physician to Queen Elizabeth 1, discovered that the earth itself was a magnet having north and south poles. The north pole of a magnet was said to point to the north pole of the earth.

16.1 The inverse square law of attraction and repulsion

In 1785 Charles Augustin de Coulomb published the fact that if two pieces of magnetic iron were placed end to end they would either attract or repel each other. Also that the amount of force between the magnets would depend on the inverse square of the distance between them.

This statement can be expressed as:

$$F \propto \frac{1}{d^2}$$

If like poles (both north or south) are placed near to each other they will repel each other, but if unlike poles (one north and the other south) are placed near to each other they will attract each other (see Fig. 16.1).

Like poles repel
Unlike poles attract

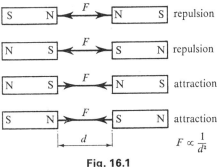

Fig. 16.1

This law conflicts with the idea that a north end of a compass needle will point to the north pole of the earth. When compasses were first used for navigational purposes this was not known and in fact did not

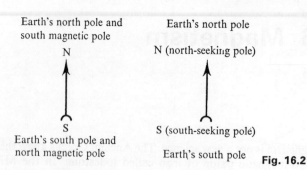

Fig. 16.2

matter, and so to avoid complications this confliction can be overcome either by saying that there is a south magnetic pole at the north pole or by saying that the magnet has a **north-seeking** pole (Fig. 16.2).

For many engineering purposes it is not necessary to consider the earth's magnetic field, and the second method is perhaps the most reasonable if the descriptions *north-seeking* pole and *south-seeking* pole are abbreviated to N pole and S pole respectively.

16.2 Field patterns

A simple experiment that can be carried out is to place a sheet of paper over a bar magnet and sprinkle iron filings onto the paper. After slight agitation a pattern will appear. A diagrammatic representation of such a pattern is shown in Fig. 16.3.

Direction of the magnetic field

Each iron filing has become a small magnet and an arrow can be inserted on some of the lines to represent the direction of the field. This direction is defined as the direction in which a unit north would move; that is, the direction in which a compass would point.

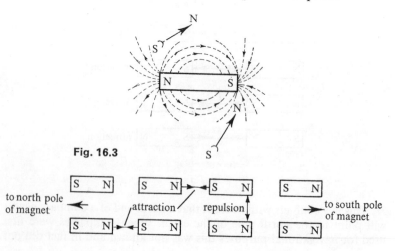

Fig. 16.3

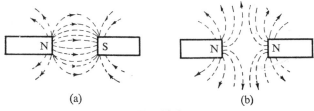

Fig. 16.4

The appearance of this pattern gives rise to the description of the field as a series of lines. The reason for the lines can be explained by the laws of magnetism, that is, repulsion and attraction. An enlarged view of a small part of the magnetic field can also be seen in Fig. 16.3.

As each iron filing has become a magnet its north and south poles will attract the south and north poles of adjacent filings and they will join end to end. Two such lines will push each other apart because of the repulsion effect between like poles.

If the experiment is repeated it will be noticed that these lines do not reappear in the same position, showing that a magnetic field is a continuous influence.

Figure 16.4(a) shows that when unlike poles are adjacent the field between the magnets is very strong, whilst Fig. 16.4(b) shows that when like poles are adjacent they produce a weak field with a null point at its centre.

16.3 Definition of magnetic field

A magnetic field may be defined as *the area surrounding a magnet in which a magnetic influence can be detected.*

Magnetic flux (Φ)

Magnetic flux is a measure of the total amount of magnetic field that is produced by a magnet. The unit of magnetic flux is the **weber** (Wb), defined in Chapter 18, Section 18.1.

Magnetic flux density (B)

Magnetic flux density is the measure of the strength of a magnetic field and is the magnetic flux within a given area:

$$B = \frac{\Phi}{A}$$

where A is the cross-sectional area of the magnetic field.

The unit used for flux density is the **tesla** (T), defined as *the density of one weber of magnetic flux per square metre.*

Example 16.1 The magnet shown in Fig. 16.5(a) has a cross-sectional area of 40 cm² and produces an effective flux of 50 mWb. The magnet

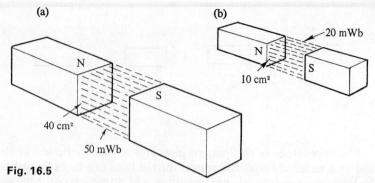

Fig. 16.5

shown in Fig. 16.5(b) has a smaller cross-sectional area of 10 cm² and produces 20 mWb of effective flux. Compare the flux density of each magnet.

The flux density of magnet (a) is

$$B = \frac{\Phi}{A} = \frac{50 \times 10^{-3}}{40 \times 10^{-4}} = 12 \cdot 5 \text{ T}$$

The flux density of magnet (b) is

$$B = \frac{\Phi}{A} = \frac{20 \times 10^{-3}}{10 \times 10^{-4}} = 20 \text{ T}$$

Thus, although more flux is produced by magnet (a) the field from (b) is more dense.

Question 16.1 A magnet of cross-sectional area 25 cm² has a flux density of 15 T. Find the total flux produced by the magnet. (37·5 mWb)

16.4 Magnetizing a piece of steel using other magnets

A piece of steel gradually becomes magnetized if it is stroked gently with a magnet.

The molecular theory of magnetism

In the mid-nineteenth century Wilhelm Eduard Weber and, later, James Clerk Maxwell proposed that each molecule within a piece of steel was itself a magnet.

Figure 16.6 shows a demagnetized piece of steel. Each arrow represents a molecule showing that by their haphazard arrangement the overall magnetic effect is nil. If the steel is magnetized, for example, by the stroking magnetizing method previously mentioned, the direction of the molecules is changed until they line up, thus producing an overall magnetic effect. This is shown diagrammatically in Fig. 16.7.

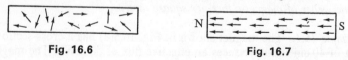

Fig. 16.6   **Fig. 16.7**

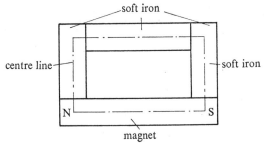

Fig. 16.8

16.5 Magnetic circuit

A magnetic circuit is the path taken by the magnetic field. For a magnet in air there would be an infinite number of such paths, as can be seen from Fig. 16.3.

If the magnetic poles are joined by pieces of soft iron, as in Fig. 16.8, then very little magnetic effect can be detected outside the iron. This arrangement is called a *closed* magnetic circuit and to open any of the joints a force must be exerted. Nearly all the magnetic flux produced by the magnet now passes through the iron circuit and the centre-line of the iron can be measured as the length of the magnetic circuit.

Reluctance (symbol S)

Reluctance is the resistance of the magnetic circuit to the presence of magnetic flux. Air and non-magnetic materials such as copper and aluminium have a very high reluctance, while irons and steels have much lower values of reluctance. Thus iron is a good magnetic conductor while air, copper, aluminium, etc., are poor magnetic conductors.

It is worth remembering the rule that **magnetic flux will try to take the path of lowest reluctance.**

Additional questions

16.2 A flux density of 100 T is produced in a core by a flux of 2 Wb. Find the cross-sectional area of the core.

16.3 A magnet produces a flux density of B tesla in an iron ring. The iron ring has an air gap which is filled by (a) a piece of copper, (b) a piece of soft iron, and (c) a piece of aluminium. Say in each case how the flux density is affected.

16.4 When bar magnets are stored they are placed as shown in Fig. 16.9, side by side with their poles joined by pieces of soft iron called keepers. Explain why you think this is done.

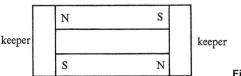

Fig. 16.9

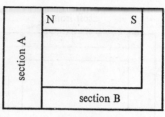

Fig. 16.10

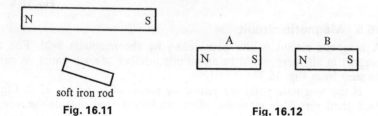

Fig. 16.11 Fig. 16.12

16.5 A magnetic circuit has a flux density of 50 T and a square cross-sectional shape of sides 2 cm. Find the total flux developed by the magnet.

16.6 Figure 16.10 shows a magnetic circuit composed of two sections. Section A has a cross-sectional area of 60 cm^2 and section B a cross-sectional area of 20 cm^2. If the permanent magnet produces a flux of 20 mWb, find the flux density in each section.

16.7 Figure 16.11 shows a soft iron rod placed into a magnetic field. Show the polarity induced into the soft iron rod and the direction of the field.

16.8 A number of metal scribers are to be stored for a considerable length of time. Will it be best to store them (a) pointing north to south or (b) east to west? Explain the reason for your answer and the result of incorrect storage.

16.9 Figure 16.12 shows two magnets A and B placed close together. Sketch the magnetic field that would result (a) if the polarity were as shown and (b) if magnet A were reversed end for end.   (C.G.L.I.)

17 Electromagnetism

The fact that the atoms forming the molecules were thought to contain electric charge caused André Marie Ampere to formulate, in 1825, the idea that magnetism was due to electric currents circulating within the molecules.

It was found that when an electric current passed through a wire, and a compass needle was placed near to the wire, the needle varied in direction whenever the current was altered; that is, when the current was started, stopped, or reversed.

Figure 17.1(a) shows the end view of a conductor with current flowing away from the viewer, and Fig. 17.1(b) shows the other end of the conductor with current coming towards the viewer.

The cross within the conductor of Fig. 17.1(a) may be thought of as the end view of an arrow or dart whilst the point within the conductor of Fig. 17.1(b) may be thought of as the pointed end of the arrow or dart.

Figure 17.2 shows how a compass needle will be deflected when positioned near to the conductor of Fig. 17.1(a).

Figure 17.3 shows the conductor viewed from the opposite end.

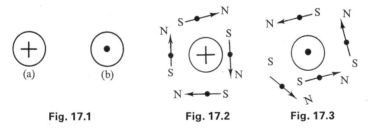

Fig. 17.1   Fig. 17.2   Fig. 17.3

These compasses form tangents to imaginary concentric circles around the conductors. The field around a current-carrying conductor is therefore shown as a series of concentric lines with arrows indicating the direction of the field.

Only the diagram shown in Fig. 17.4 needs to be remembered. This is best done by thinking of a normal, right-handed type of cross-head screw. To screw inwards, in the direction of the current, the screw would be turned in a clockwise direction, which is the direction in which the compass will point.

As Fig. 17.5 shows the screw coming out of the conductor, the direction of the field must be reversed.

Fig.17.4

Fig. 17.5

17.1 Field produced by two parallel conductors

Currents in the same direction

Consider a compass needle placed between the two conductors in Fig. 17.6. One field will try to make the needle point upwards while the other will try to point the needle downwards.

Both fields are thus pushed away from the centre and the two conductors are surrounded by a single field. This will be as shown in Fig. 17.7.

If the currents are both reversed the field pattern will be the same but the direction of the field will be reversed. As the magnetic flux will try to take the path of lowest reluctance, the field will try to bring the conductors together in order to shorten its magnetic circuit, thus exerting a force on the conductors.

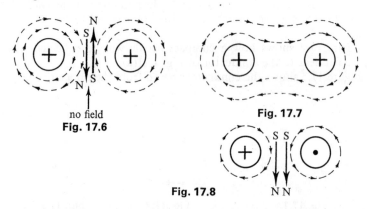

Currents in opposite directions

If two magnetic needles are placed side by side between the conductors (Fig. 17.8), both will point in the same direction. All the magnetic flux produced by the conductors will pass between the conductors, and the field will be dense.

Question 17.1 Draw the field pattern that would result from parallel conductors having currents flowing in them in opposite directions.

Owing to the law of repulsion between like poles, the field will try to open out and will thus cause a force to be exerted between the conductors.

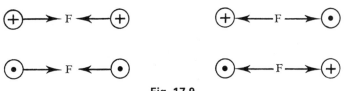

Fig. 17.9

Figure 17.9 summarizes the force exerted on parallel current-carrying conductors.

17.2 Field from a solenoid

A **solenoid** is a coil of wire whose length is large in comparison with its diameter. A sectional view of a solenoid can be seen in Fig. 17.10.

The field will be seen to be a combination of the effects of Fig. 17.6 and Fig. 17.8, and the general field pattern will be seen to be similar to that of a bar magnet (see Fig. 16.3).

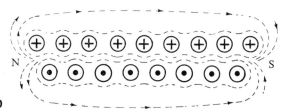

Fig. 17.10

17.3 Magnetomotive force (*F*)

Magnetomotive force (m.m.f.) is the product of the current and the number of turns of conductor wire:

$$F = NI$$

The unit used is the ampere-turn (At), which may be shortened to A if this cannot be confused with the symbol for current. However, this is not recommended at this stage.

Example 17.1 A solenoid is wound with 800 turns of wire and a current of 3 A is passed through the wire. Find the magnetomotive force.

$$F = NI = 800 \times 3 = 2400 \text{ At} \quad \text{or} \quad \text{A}$$

Question 17.2 A magnetomotive force of 1000 At is required from a coil having 2000 turns of wire. How much current must be passed through the wire? (0·5 A)

17.4 The 'Ohm's law' of the magnetic circuit

The ratio of magnetomotive force to magnetic flux for a magnetic circuit is called 'reluctance'. In Chapter 16, Section 16.5, reluctance was defined as *the resistance of the magnetic circuit to the presence of magnetic flux*. The symbol for reluctance is *S*.

$$\text{Reluctance} = \frac{\text{Magnetomotive force}}{\text{Magnetic flux}}$$

$$S = \frac{F \text{ (ampere-turns)}}{\Phi \text{ (webers)}}$$

From this formula a unit of reluctance, the ampere-turn per weber (At/Wb), is derived. (A/Wb may also be used.)* This formula for reluctance may be compared with the Ohm's law formula for resistance in a d.c. circuit ($R = E/I$).

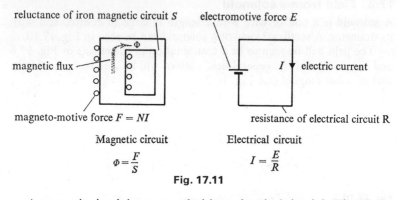

Fig. 17.11

A magnetic circuit is compared with an electrical circuit in Fig. 17.11.

From this comparison it will be seen that the m.m.f. is compared with the e.m.f. of the electrical circuit. The product of the current and number of turns on a coil, the ampere-turns, is the prime cause of the resulting magnetic field.

If a soft iron core is inserted into the centre of the solenoid of Fig. 17.10, the magnetic effect will be greater since the reluctance of soft iron is much lower than the reluctance of air.

From the formula $S = F/\Phi$, and rearranging to $\Phi = F/S$, if the m.m.f. F is unchanged a reduction in S will increase Φ.

Example 17.2 A magnetic circuit has a reluctance of 5×10^6 At/Wb. Find the current which is required to flow in a solenoid of 10 000 turns of wire if the desired level of flux is 5 mWb.

$$\Phi = \frac{F}{S}$$

or $\qquad F = \Phi S = 5 \times 10^{-3} \times 5 \times 10^6 = 25 \times 10^3$ At

also $\qquad F = NI$

or $\qquad I = \frac{F}{N} = \frac{25 \times 10^3}{10 \times 10^3} = 2\cdot 5$ A

* A unit for reluctance, 1/H (reciprocal henry) can also be developed (see Volume II, Chapter 5).

Question 17.3 If the materials comprising the magnetic circuit of Example 17.2 are changed so that the reluctance is reduced to 4×10^4 At/Wb, to what number may the turns be reduced if the magnetic flux and electric current are to be unchanged. (80 turns)

Magnetic field intensity (symbol H)

The magnetic field intensity is the magnetomotive force per unit length of magnetic circuit:

$$H = \frac{F}{l}$$

As the unit of magnetomotive force, F, is the ampere-turn, and the unit of length is the metre, the unit of magnetic field intensity will be

$$H = \frac{F \text{ (ampere-turns)}}{l \text{ (metres)}}$$

that is, ampere-turns per metre (At/m). (The unit A/m may also be used.)

Care must be taken to use the correct value of length which is the *length of magnetic circuit* and **not** the length of wire on the solenoid.

Example 17.3 A coil of 80 turns of wire is wound on an iron ring having a mean diameter of 5 cm. Find the magnetic field intensity produced when a current of 10 A is passing through the coil.

$$F = NI = 80 \times 10 = 800 \text{ At}$$

The length of the magnetic circuit is

$$\pi d = 3 \cdot 142 \times 5 \times 10^{-2} = 0 \cdot 1571 \text{ m}$$

$$H = \frac{F}{l} = \frac{800}{0 \cdot 1571} = 5100 \text{ At/m}$$

Question 17.4 The magnetic field intensity produced from a coil of wire wound on an iron ring of 8 cm mean diameter is 4000 At/m. If the current in the coil is 5 A, find the number of turns needed for the winding. (201)

17.5 Absolute permeability (μ)

Absolute permeability can be considered as the factor that links the electric current with the magnetic field caused by that current.

The flux density, which is the measure of magnetic effect, is equal to the product of absolute permeability and the magnetic field intensity, the latter being the measure of the electrical input:

$$B = \mu H$$

The unit of absolute permeability is the henry per metre (H/m). The unit of inductance, the henry, will be discussed in Chapter 18.

From the formula, $B = \Phi/A$, flux density can be seen to be a magnetic effect; that is, it is magnetic flux and a magnetic circuit constant.

In a similar way the formula $H = F/l = NI/l$ shows that magnetic field intensity is an electrical effect, being comprised of electric current and magnetic circuit constants.

Thus magnetic flux is proportional to the product of absolute permeability and the electric current.

Absolute permeability is divided into two parts: the *permeability of free space* and the *relative permeability*.

Permeability of free space (μ_0)

The permeability of free space is a constant for all materials and has a value of $4\pi \times 10^{-7}$ H/m.

Relative permeability (μ_r)

The relative permeability is a variable factor. For non-magnetic materials such as air, copper, aluminium, rubber, etc., the relative permeability has a value of unity.

For magnetic materials, irons and steels, the value will vary—typical values being between 100 and 2500, although special alloys can produce much higher values. Relative permeability is a factor and has no units.

Absolute permeability is found from the product of permeability of free space and the relative permeability:

$$\mu = \mu_0 \times \mu_r$$
$$= 4\pi \times 10^{-7} \times \mu_r \text{ H/m}$$

Example 17.4 A current through an air spaced coil produces a magnetic field intensity of 600 At/m. Find the flux density of the magnetic field.

As the coil is air spaced $\mu_r = 1$.

$$\mu = \mu_0 \times \mu_r$$
$$= 4\pi \times 10^{-7} \times 1 \text{ H/m}$$
$$B = \mu H = 4\pi \times 10^{-7} \times 600 \text{ T}$$
$$= 754 \cdot 3 \times 10^{-6} \text{ T} \quad \text{or} \quad 754 \cdot 3 \text{ } \mu\text{T}$$

Question 17.5 The magnetic field intensity produced by a solenoid is 1800 At/m. If the solenoid has a magnetic circuit constructed of a material having a relative permeability of 400, find the flux density in the magnetic circuit.

(0·9054 T)

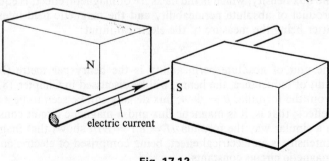

Fig. 17.12

Using the magnetic formulae introduced so far, an alternative formula for reluctance may be deduced:

$$\Phi = BA = \mu HA = \frac{\mu FA}{l}$$

also

$$\Phi = \frac{F}{S}$$

therefore

$$\frac{F}{S} = \frac{\mu FA}{l}$$

$$\frac{1}{S} = \frac{\mu A}{l}$$

and

$$S = \frac{l}{\mu A}$$

Example 17.5 A cast iron ring of 15 cm mean diameter has a cross-sectional area of 4 cm² and a relative permeability of 1500. Find the reluctance of the ring.

$$S = \frac{l}{\mu_0 \mu_r A}$$

$$= \frac{\pi \times 15 \times 10^{-2}}{4\pi \times 10^{-7} \times 1500 \times 4 \times 10^{-4}}$$

$$= 0.625 \times 10^6 \text{ At/Wb}$$

Question 17.6 A transformer has a magnetic circuit length of 40 cm and cross-sectional area of 10 cm². If the steel used has a relative permeability of 1200, find the reluctance of the magnetic circuit.   (0.265×10^6 At/Wb)

17.6 Types of magnetic materials

Magnetic materials can be divided into two groups:
  (1) Those which have a high relative permeability and can be magnetized easily. These materials are used, for example, in transformers, relays, motors, and generators.
  (2) Those which will retain a high level of magnetism and can be used as permanent magnets. These materials are used, for example, in loudspeakers, telephone receivers, and measuring instruments.

Iron, cobalt, and nickel are the most common ferromagnetic elements, and these are mostly used in alloys.

Examples of magnetic alloys are silicon steels, or electrical sheet steel, which is iron plus 3–4 per cent silicon, and permalloys consisting of nickel and iron.

17.7 Force on a conductor in a magnetic field

Consider the conductor shown in Fig. 17.12 positioned at right angles to a permanent magnetic field.

The effect of the magnetic field on the current-carrying conductor is best considered by sketching a sectional view of the conductor as shown in Fig. 17.13 and considering the permanent and electromagnetic fields separately.

When the two fields are brought together, the direction of the fields *above* the conductor are the same, but *below* the conductor the fields are in opposition. Above the conductor the two fields combine to form a strong magnetic field (see Fig. 17.14) and below the conductor the two fields reject each other and the field becomes non-existent.

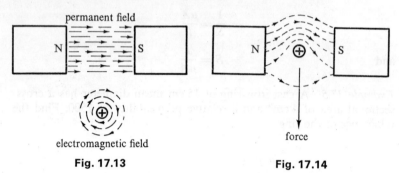

Fig. 17.13   **Fig. 17.14**

As the magnetic circuit through the air has been lengthened, and as the magnetic flux tries to take the path of lowest reluctance, which would be straight between the poles, the field will try to remove the conductor causing a force to act on the conductor in a downward direction.

This effect of force on a conductor due to the interaction of two magnetic fields is the principle upon which the operation of electric motors depends.

The magnitude of the force will depend on the length of the conductor in the permanent field, the density of the permanent field, and the value of electric current causing the electromagnetic field.

For a single conductor at right angles to a permanent field, the force is given by

$$F = BIl$$

where F is the force on the conductor in newtons, B is the flux density in tesla, and l is the length of conductor within the permanent magnetic field in metres.

For more than one conductor, for example N conductors, the force will be

$$F = BIl \times N \text{ newtons}$$

Example 17.6 A cable consisting of five conductors is situated at right angles to a magnetic field of density 3 T. If the current in each conductor is 10 A, all currents being in the same direction, and if the effective length of the cable in the field is 10 cm, find the force on the cable.

$$F = BIlN \text{ newtons}$$
$$= 3 \times 10 \times 10 \times 10^{-2} \times 5$$
$$= 15 \text{ N}$$

Question 17.7 If the current in the cable of Example 17.6 is increased to 15 A, find the force on the cable. (22·5 N)

Question 17.8 Two of the currents are now reversed. How is the force changed? (4·5 N)

17.8 Definition of the ampere

Figure 17.9 illustrated the force acting on parallel conductors carrying electric currents. In the SI system the unit of current, the ampere, is defined using this effect.

The unit of electric current called the 'ampere' is that constant current which, if maintained in two parallel rectilinear (straight) conductors of infinite length, of negligible circular cross-section, and placed at a distance of one metre apart in a vacuum, would produce between these conductors a force equal to 2×10^{-7} *newton per metre length* (B.S. 3763).

The essential details of this definition can be remembered by a sketch such as that shown in Fig. 17.15.

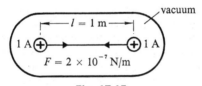

Fig. 17.15

The operation of the moving coil instrument, both ammeter and voltmeter, also depends upon the force caused when a conductor carrying an electric current is situated at right angles to a magnetic field. The principle of the moving coil instrument and the d.c. electric motor will be dealt with in Volume II.

Additional questions

17.9 If a coil is wound with 3000 turns of wire, find the current required to produce an m.m.f. of 15 At.

17.10 If the coil of Question 17.9 is wound on a soft iron ring of diameter 8 cm, find the magnetic field intensity.

17.11 An air spaced solenoid, that is a solenoid wound on a non-magnetic hollow former, produces a magnetic field intensity of 250 At/m. Find the flux density of the magnetic field.

17.12 If the solenoid of Question 17.11 has a soft iron core with a relative permeability of 500 inserted into its centre, what effect will there be on the flux density of the magnetic field produced.

17.13 The flux density produced in a round sectioned piece of soft iron, having a diameter of 2 cm, is 8 T. Find the total flux in the iron.

17.14 An electromagnetic relay has a magnetic circuit length of 30 cm and an average cross-sectional area of 1·5 cm². When the magnetic circuit of the relay is closed by the operation of the armature, the magnetic material forming the magnetic circuit has a relative permeability of 850. Calculate the reluctance of the magnetic circuit.

17.15 A current of 500 mA is passed through a coil having 1500 turns of wire. If the reluctance of the magnetic circuit is 8×10^6 At/Wb, find the total flux produced.

17.16 A solenoid, having 200 turns of wire, develops a magnetic field intensity of 1500 At/m when a current of 3 A is flowing in the wire. If the iron of the magnetic circuit has a relative permeability of 800, find (a) the length of the magnetic circuit and (b) the flux density of the field produced in the iron.

17.17 A current of 0·5 A is passed through a coil of 800 turns of wire wound onto an iron ring of mean diameter 15 cm. If the relative permeability of the iron is 550 and the diameter of the round cross-section of the iron ring is 3 cm, find the flux developed in the core.

17.18 Two vertical straight conductors are placed a short distance apart. Sketch, in plan view, the magnetic field when the current in both conductors flows (a) upwards in both and (b) in opposite directions. State in each case whether the conductors will attract or repel one another. (C.G.L.I.)

17.19 A straight current-carrying conductor is placed in a uniform magnetic field at right angles to the lines of force. Make a sketch of the resulting magnetic field, showing clearly the directions of current and magnetic flux. Show also the direction of the force on the conductor. (C.G.L.I.)

17.20 In Fig. 17.16 abcd is a rectangular coil wound with 50 turns of wire in which a current of 0·5 A flows in the direction shown. The coil is in a magnetic field of uniform flux density $B = 0.25$ T as indicated.

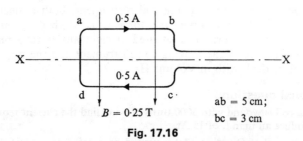

Fig. 17.16

(a) Find the force acting on each of the sides of the coil ab and cd, giving the magnitude and direction in both cases.
(b) If the coil were free to rotate about axis *XX*, find the torque tending to turn it and say how far it would turn before the torque became zero, giving reasons. (C.G.L.I.)

17.21 A conductor carrying a current of 5 A and having active length 9 cm lies in and at right angles to a uniform magnetic field. If the force on the conductor is 100 gf find the strength of the magnetic field, stating its units. (C.G.L.I.)

18 Electromagnetic Induction

18.1 The generation of an electromotive force by a permanent magnet and coil

Figure 18.1 shows a permanent magnet about to be moved into a coil of wire, the ends of which are connected to a galvanometer.

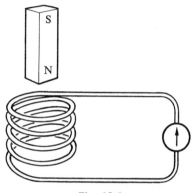

Fig. 18.1

The **galvanometer** is an ammeter which reads forwards and backwards from a central zero position, and is therefore useful for reading currents in either a forward or reverse direction without a change of connection.

On the insertion of the magnet the galvanometer will deflect, and on the magnet's removal the galvanometer will again deflect but in the opposite direction.

Any current that flows will only do so if an e.m.f. is present. The e.m.f caused by the insertion of the magnet into the coil is termed an **induced electromotive force.** A current will only be caused by the induced e.m.f. if a circuit exists in which the current can flow.

The direction of induced current

When the permanent magnet is moved towards the coil shown in Fig. 18.1, the induced current in the coil will cause a force that will *oppose* the movement. This opposing force must be caused by an induced north pole at the top of the coil. The effect will be similar to the bringing together of like permanent magnetic poles, as shown in Fig. 18.2.

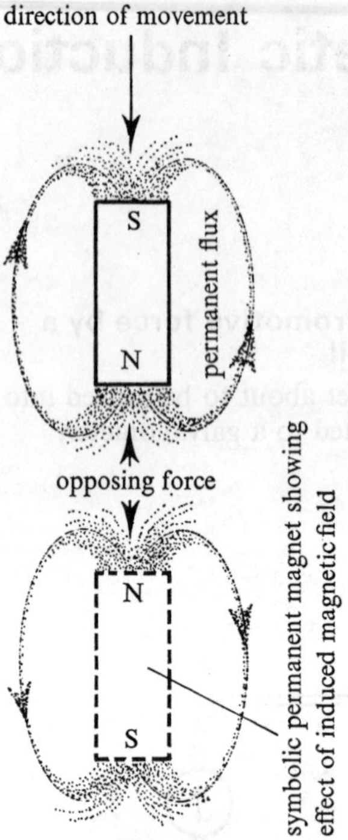

Fig. 18.2

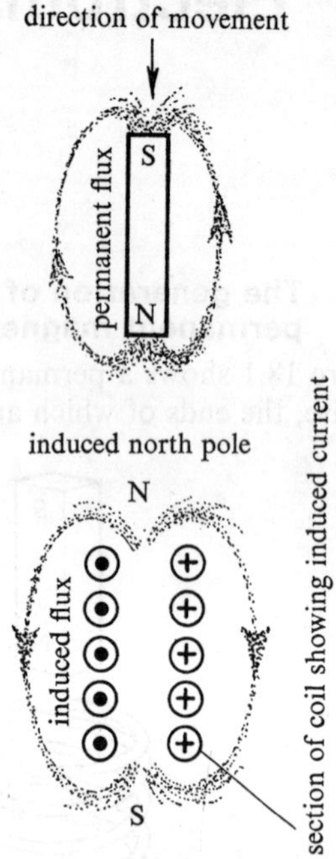

Fig. 18.3

The direction of the induced current must therefore be similar to that shown in Fig. 18.3.

On removal of the magnet a force will be felt and this must be due to the coil attracting the magnet as shown in the explanatory diagram, Fig. 18.4.

The direction of the induced current will be as shown in Fig. 18.5, and will be seen to be the reverse of that shown in Fig. 18.3.

The direction of the current may either be determined as shown in Chapter 17, Section 17.2, or by placing the right hand round the coil with the thumb extended in the direction of the north pole, when the other fingers will indicate the direction of the current.

Magnitude of the induced electromotive force: Faraday's law

If the permanent magnet is moved slowly the deflection on the galvanometer will also be small. If the speed of the movement is increased the deflection will also increase. Therefore,

Induced electromotive force is proportional to the velocity of the magnetic field movement.

If another coil, having a larger number of turns of wire but with the same electrical resistance, is substituted for the coil of Fig. 18.1 the deflection will be increased. Therefore,

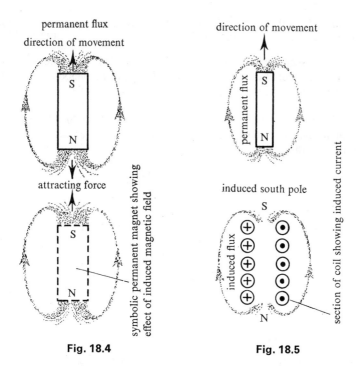

Fig. 18.4   **Fig. 18.5**

Induced electromotive force is proportional to the number of turns of wire on the coil.

If a 'stronger' magnet is used the deflection will again increase. Faraday joined these variations together in the form of a law:

$$E \propto \frac{1}{\text{Time of movement}}$$
$$E \propto \text{Number of turns}$$
$$E \propto \text{Magnetic flux}$$
$$E = -N\frac{\Delta\Phi}{\Delta t} \text{ volts}$$

where the negative sign shows that the current will oppose the movement, N stands for the number of turns on the coil, and $\Delta\Phi/\Delta t$ is the average rate of change of the flux.

Rate of change of flux

Example 18.1 A flux of 10 mWb passes across a coil in a time of 3 ms. Find the average rate of change of flux.

$$\frac{\Delta\Phi}{\Delta t} = \frac{10 \times 10^{-3}}{3 \times 10^{-3}} = 3{\cdot}33 \text{ Wb/s}$$

GENERATION OF AN ELECTROMOTIVE FORCE

Question 18.1 A flux of 500 Wb takes a time of 10 min to pass across a coil. Find the average rate of change of flux. (0·833 Wb/s)

It will be seen from the answers to Examples 18.1 and Question 18.1 that it is the combination of the size of field and the velocity of movement that determine the value of the rate of change of flux.

Example 18.2 A flux of 0·3 Wb passes across a coil of 200 turns of wire in a time of 0·6 s. Find the average induced e.m.f. across the terminals of the coil.

$$E = -N\frac{\Delta\Phi}{\Delta t} = -200 \times \frac{0\cdot3}{0\cdot6} = -100 \text{ V}$$

The induced e.m.f. will be 100 V in a direction that would tend to produce a current opposing the movement of the field.

Question 18.2 A flux of 200 mWb takes a time of 0·08 s to pass across a coil of 500 turns of wire. Find the average induced e.m.f. across the coil.
(−1250 V (in opposition))

Definition of the unit of magnetic flux

From Faraday's law, $E = -N(\Delta\Phi/\Delta t)$, the unit of magnetic flux, the **weber,** may be defined:

The unit of magnetic flux called the weber is the flux which linking a circuit of one turn produces in it an electromotive force of one volt as it is reduced to zero at a uniform rate in one second (B.S. 3763).

This definition puts unity values into the formula $E = -N(\Delta\Phi/\Delta t)$:

$$1 \text{ (volt)} = -1 \text{ (turn)} \times \frac{1 \text{ (weber change)}}{1 \text{ (second taken)}}$$

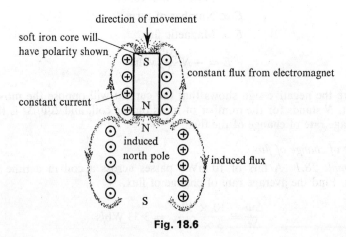

Fig. 18.6

Magnitude of the induced current

The value of induced current will depend upon the induced e.m.f. and the resistance of the circuit.

Example 18.3 If the coil in Example 18.2 has a resistance of 20 Ω and an ammeter of resistance 5 Ω is connected across the terminals of the coil, find the induced current in the circuit when the conditions are as stated in Example 18.2.

$$\text{Induced current} = \frac{\text{Induced e.m.f.}}{\text{Total circuit resistance}}$$
$$= \frac{-100}{20 + 5} = -4 \text{ A}$$

The induced current will be 4 A in a direction such as to cause a force opposing the movement.

Question 18.3 Find the induced current in the coil of Question 18.2 if the resistance of the coil is 1000 Ω and an ammeter of resistance 10 Ω is connected across the coil terminals. The other conditions are as Question 18.2.

$$(-1.24 \text{ A (in opposition)})$$

Effect of replacing the permanent magnet by an electromagnet

If instead of a permanent magnet an electromagnet is used, then providing that the current through the electromagnet coil is kept constant, the effect will be unchanged.

Figure 18.6 shows the similarity of the arrangement when compared with Fig. 18.3.

If the electromagnet is inserted into the coil, the induced e.m.f. will fall to zero when the movement has ceased. If the current in the electromagnet is changed a deflection will be noticed on the galvanometer but only *while the current is changing*.

This arrangement is shown in Fig. 18.7.

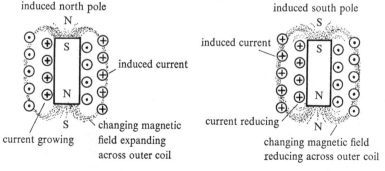

Fig. 18.7 Effect of increasing current in centre coil

Fig. 18.8 Effect of reducing current in centre coil

GENERATION OF AN ELECTROMAGNETIC FORCE

The cause of this deflection is as follows:

If the current in the centre coil is increasing in magnitude, the flux developed by the coil will increase and the field will expand, crossing the outer coil as it does so. An e.m.f. will be induced into the outer coil in such a direction that it will tend to cause a field that will oppose the change; that is, a field of reverse polarity.

If the current in the centre coil is reducing in magnitude, as shown by Fig. 18.8, the flux developed by the coil will reduce and the field will collapse. The field will now cross the outer coil in the opposite direction and the direction of the induced e.m.f. will be reversed. The field caused by this induced e.m.f. will oppose the change; that is, it will attempt to continue the original direction of the flux.

Thus, an e.m.f. may be induced into one coil by a current changing in another coil if the two coils are magnetically coupled.

18.2 Mutual induction

When an electromotive force is induced in a circuit due to the current changing in another circuit, the effect is termed **mutual induction.**

Mutual inductance

When induction exists between two circuits they are said to have the property of **mutual inductance.** The symbol used for mutual inductance is M and its unit is the henry (H).

18.3 Self-induction

When a current changes in value in a coil, as shown by Fig. 18.9, the growing or falling magnetic field caused by each and every turn of the coil will pass across the remainder of the coil.

The effect of this must be similar to the effects shown by Figs. 18.7 and 18.8.

Figure 18.10 shows the effect of a growing current. In a similar way to Fig. 18.7 a current will be induced in opposition to the main current.

Figure 18.11 is an explanatory diagram showing the induced e.m.f. as a generator connected, with opposing polarity, in series with the

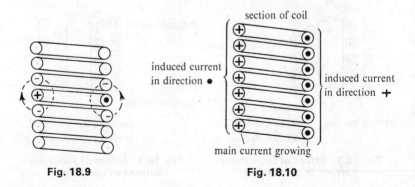

Fig. 18.9   Fig. 18.10

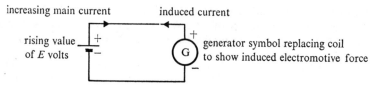

Fig. 18.11 Explanatory diagram

battery causing the main current. The explanatory generator will only be in the circuit whilst the current is growing. The effect of this induced e.m.f., often termed a 'back e.m.f.', will be to slow down the growth of the main current.

Figure 18.12 shows the effect of a reduction in current.

The induced current in this case will be similar to Fig. 18.8, and will be in a direction which will attempt to prevent the fall in the main current.

Figure 18.13 is an explanatory diagram of a generator replacing the

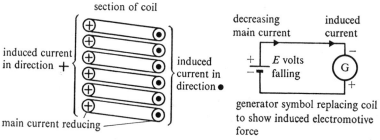

Fig. 18.12   **Fig. 18.13 Explanatory diagram**

induced e.m.f. In this case the generator is aiding the battery causing the main current.

Self-inductance

When a circuit is self-inductive it is said to have the property of **self-inductance**. The symbol for self-inductance is L and its unit is the henry, which is also the unit for mutual inductance. The henry is defined as follows:

The unit of electric inductance called the henry is the inductance of a closed circuit in which an electromotive force of one volt is produced when the electric current in the circuit varies uniformly at the rate of one ampere per second (B.S. 3763).

18.4 Lenz's law

Lenz's law is a concise statement of the effects described so far:

An induced electromotive force will be in a direction such that any current that it may cause will oppose the movement or change of magnetic flux responsible for inducing that electromotive force.

Example 18.4 A steady current of 1 A flowing in a coil of 60 turns of wire causes a magnetic flux of 30 mWb. If the current is reversed in a time of 20 ms, find the self-induced e.m.f. developed across the coil.

Current changes from 1 A to −1 A, which is a change of 2 A. The change of flux will be 30 mWb to −30 mWb, a change of 60 mWb.

$$E = -N\frac{\Delta\Phi}{\Delta t} = -60 \times \frac{60 \times 10^{-3}}{20 \times 10^{-3}} = -60 \times 3$$
$$= -180 \text{ V}$$

The induced e.m.f. is 180 V in a direction opposite to the change of applied e.m.f.

Question 18.4 A steady current of 250 mA in a coil is found to cause a flux of 800 μWb. The current is reversed in a time of 2·5 ms. Find the induced e.m.f. developed across the coil if the coil has 1000 turns of wire. Comment on the direction of this e.m.f. (640 V (in opposition))

Example 18.5 If the current in the coil of Example 18.4 is reduced from 1 A to zero in a time of 2 ms, find the induced e.m.f. (Assume that the flux is proportional to the current.)

If the current falls from 1 A to 0, then the flux falls from 30 mWb to 0:

$$E = -N\frac{\Delta\Phi}{\Delta t} = -60 \times \frac{30 \times 10^{-3}}{2 \times 10^{-3}} = -900 \text{ V}$$

The e.m.f. is 900 V in a direction which will try to continue the flow of the current.

Question 18.5 If the current in the coil of Question 18.4 is halved in a time of 200 μs, find the induced e.m.f. (Assume that the flux is proportional to the current.) (2000 V (in opposition))

18.5 Practical applications of induced electromotive force

Figure 18.14 shows the coil of Example 18.5 connected to a d.c. supply via a switch.

When the switch is 'broken' in a time of 2 ms a p.d. of 900 V will develop across the coil. This p.d. may well be sufficient to cause the

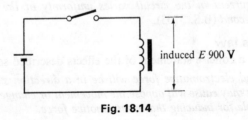

Fig. 18.14

insulating air gap across the switch blades to break down and an arc will then occur.

Another example of an induced e.m.f. causing an induced current is termed an **eddy current.**

Figure 18.15 shows a simplified view of the magnets and coil of a moving coil meter.

When a current is passed through the coil it will be caused to rotate by force F_1. As the coil starts to move across the magnetic field a current will be induced into the coil former which is normally made of aluminium. From Lenz's law this current will cause a reverse force, which is shown in Fig. 18.16 as F_2.

The force now causing the coil to rotate will be $F_1 - F_2$. As F_2 depends on the velocity of the coil, this force will fall to zero as the instrument needle reaches its reading point. The needle will be prevented from oscillating and is said to be 'damped'.

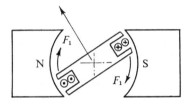

Fig. 18.15   Fig. 18.16

Additional questions

18.6 A coil of 800 turns of wire has a current flowing of 2 A. If this current causes a flux of 650 μWb, find the induced e.m.f. when the current is reversed in a time of 50 ms. Explain the direction of this induced e.m.f.

18.7 A single loop of wire is passed across a field of 100 mWb in a time of 4 ms. Find the e.m.f. induced in the wire and comment on its direction.

18.8 The aluminium former of a moving coil ammeter crosses a flux of 40 μWb in a time of 1 ms. If the resistance of the former is 4×10^{-3} Ω, find the induced current in the former.

18.9 A p.d. of 200 V is induced in a coil of 800 turns when the coil is passed through a field of 20 μWb. Find the time taken.

18.10 State Lenz's law of electromagnetic induction.

Why may arcing occur between the contacts of a switch used to break the current in a low-voltage circuit containing an electromagnet?

With the aid of diagrams, describe how mutual inductance can be demonstrated.   (C.G.L.I.)

18.11 Explain what is meant by electromagnetic induction, and state one rule which can be used to deduce the direction of an induced e.m.f. Show how this can be used to find the direction of the e.m.f. induced in the right-hand coil of Fig. 18.17 if the current in the other coil is increasing.

Describe and explain one practical example of electromagnetic damping. (C.G.L.I.)

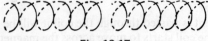

Fig. 18.17

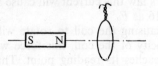

Fig. 18.18

18.12 State Lenz's and Faraday's laws of electromagnetic induction.

A bar magnet is placed on the axis of a circular coil as shown in Fig. 18.18, and is moved towards the coil. Explain how the direction of the induced e.m.f. can be predicted.

State, with reasons, the effect on the induced e.m.f. of (a) repeating with the magnet reversed end to end, (b) using a coil with three times as many turns, (c) moving the magnet at half the previous speed, (d) moving the coil as well as the magnet in such a way that it stays at the same distance from the magnet, and (e) short circuiting the coil.

(C.G.L.I.)

19 An Introduction to Alternating Current, the Generator and Motor Effect

19.1 Rotation of a wire loop in a constant, parallel magnetic field

Figure 19.1 shows a continuous loop of conductor wire, AB, being rotated about axis X–X. The loop at all times is within a constant and parallel magnetic field. Figure 19.2 is a sectional view of the loop and magnetic field shown in Fig. 19.1.

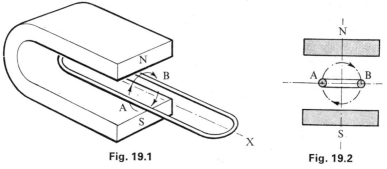

Fig. 19.1   Fig. 19.2

To enable the effect on conductor A to be considered in isolation, Fig. 19.3 shows a sectional view of that conductor only within the field.

In the position shown in Fig. 19.3, conductor A will be seen to be moving in the same direction as the field. Since, at this instant, the conductor is not crossing the field, the rate of change of field is zero and thus the induced e.m.f. is also zero. (See Faraday's law, $E = -N(\Delta\Phi/\Delta t)$, Chapter 18, Section 18.1.)

Fig. 19.3   Rotation 0°   Fig. 19.4   Rotation 90°

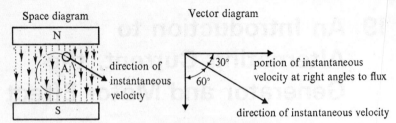

Fig. 19.5 Rotation 120°

When conductor A has turned through an angle of 90°, so shown by Fig. 19.4, its direction of movement will be at right angles to the magnetic field.

The rate at which the conductor is crossing the field will now be a maximum, and so the e.m.f. induced into the conductor will also be a maximum value. Let the angular velocity of the loop be such that the e.m.f. induced in this position equals 10 V.

Figure 19.5 shows conductor A when it has rotated through an angle of 120°. The velocity of the conductor can be shown as a vector and drawn to scale. This vector may be resolved into right angle components, one at 90° to the flux and the other in the same direction as the flux. This is shown by the vector diagram of Fig. 19.5.

The component of velocity at right angles to the flux, when measured, will be found to be 0·866 of the actual velocity. The induced e.m.f. will thus be 0·866 of the maximum value, that is $0·866 \times 10 = 8·66$ V.

Question 19.1 Find, by drawing a vector diagram of velocity, the induced e.m.f. when the conductor has rotated to 150°. (5 V)

If vector diagrams are drawn for various positions of conductor A, the result will be found to be as follows:

Table 19.1

Angle of rotation, degrees	Induced e.m.f., volts	Angle of rotation, degrees	Induced e.m.f., volts
0	0	210	−5
30	5	240	−8·66
60	8·66	270	−10
90	10	300	−8·66
120	8·66	330	−5
150	5	360	0
180	0		

Between 180° and 360° the induced e.m.f. is reversed in direction, as the direction of movement of the conductor relative to the field has reversed.

If the results shown in Table 19.1 are plotted, the shape of the graph will be as shown in Fig. 19.6.

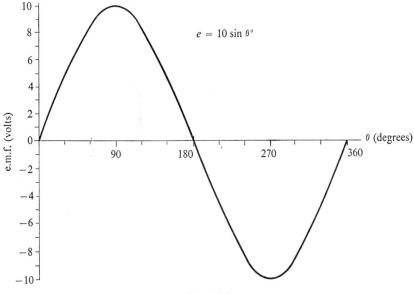

Fig. 19.6

This shape of graph is called a sinusoidal waveform; that is, it follows a sine law.

Using the symbols y and x, this law would be $y = 10 \sin x°$ or, in a more practical form, $e = 10 \sin \theta°$ where e is the instantaneous value of the induced e.m.f., θ is the angle of rotation, and the 10 is the maximum value of the waveform.

Example 19.1 Find the instantaneous value of the induced e.m.f. when the waveform obeys the law $e = 10 \sin \theta°$ and θ has a value of 60°.

$$e = 10 \sin \theta° = 10 \times \sin 60° = 10 \times 0 \cdot 866 = 8 \cdot 66 \text{ V}$$

This answer can be checked from Table 19.1.

Question 19.2 Check, by using the law $e = 10 \sin \theta°$, the values in Table 19.1 for $\theta = 0°$, 30°, and 90°. (0, 5 V, 10 V)

Rotating field system

If the conductor loop is stationary and the magnetic field is rotated, as

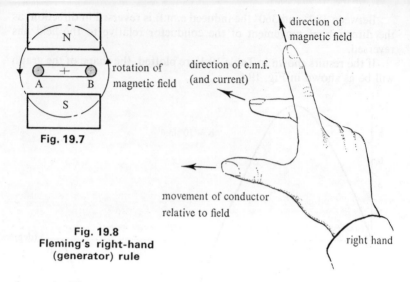

Fig. 19.7

Fig. 19.8
Fleming's right-hand (generator) rule

shown in Fig. 19.7, the relative movement of field and conductor remains unchanged, as will the values of the induced e.m.f.

The generated waveform will again be as Fig. 19.6.

The direction of the induced electromotive force

The direction of the induced e.m.f. may be obtained by the use of **Fleming's right-hand rule.**

The thumb, first, and second fingers of the right hand are held at right angles to each other. As shown in Fig. 19.8 they will then represent, respectively, the directions of conductor movement relative to the magnetic flux, the direction of the field, and the direction in which the generated e.m.f. will tend to send a current.

Example 19.2 Find the direction of the current flowing in conductor A, Fig. 19.4.

By applying Fleming's right-hand rule the direction of the current will be away from the viewer.

Describing the sine wave

Figure 19.9 shows a sinusoidal waveform representing an alternating e.m.f.

Cycle. A cycle of e.m.f. is one complete section of the waveform taken between any two similar points. Examples of these are shown on Fig. 19.9.

Frequency. The frequency of the waveform is the number of cycles of the waveform completed in 1 s. The symbol used is f and the unit in which it is measured is the hertz (Hz).

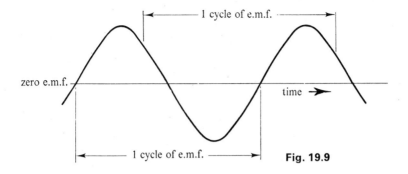

Fig. 19.9

Period. The period of a waveform is the time taken to complete one cycle. The symbol used is τ and the unit is the second.

The frequency and period of a waveform are connected by the formula

$$f = \frac{1}{\tau} \quad \text{or} \quad \tau = \frac{1}{f}$$

Example 19.3 If the frequency of a supply is 50 Hz find the time taken for one cycle.

$$\text{Time for 1 cycle (or period)} = \frac{1}{f} = \frac{1}{50} \text{ s}$$

$$\text{or} = \frac{1000}{50} = 20 \text{ ms}$$

Question 19.3 If the periodic time of a waveform is 0·1 ms, find the supply frequency. (10 kHz)

Peak value. The peak value of a waveform is the highest or maximum positive or negative value.

Example 19.4 Draw a graph of 1 cycle of a sinusoidal e.m.f. if its frequency is 200 Hz and its peak value is 4 V.

The axes will be e.m.f. and time.

$$\tau = \frac{1}{f} = \frac{1}{200} \text{ s} = \frac{1000}{200} = 5 \text{ ms}$$

Figure 19.10 shows the waveform.

Question 19.4 Draw, using scales of e.m.f. against time, a graph of 2 cycles of a sinusoidal waveform that has a periodic time of 40 ms and a peak value of 15 V. Find the frequency of the waveform. (25 Hz)

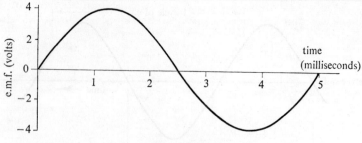

Fig. 19.10

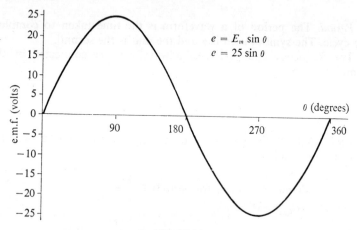

$e = E_m \sin \theta$
$e = 25 \sin \theta$

Fig. 19.11

Example 19.5 Sketch the waveform $e = 25 \sin \theta$ (volts) for 1 cycle.

The 25 is the peak value of the waveform and as e is given in volts the peak value is 25 V.

Figure 19.11 shows the waveform.

The general law of this type of waveform is thus $e = E_m \sin \theta$, where e is the instantaneous value and E_m the maximum or peak value of the e.m.f.

Question 19.5 Plot, using scales of e.m.f. and angle of rotation, the graph of the waveform $e = 75 \sin \theta$ (millivolts) for 2 cycles. From the graph find the value of the e.m.f. after a rotation angle of 390°. (37·5 mV)

Application of an alternating potential difference across a pure resistor

A pure resistor is one which has neither an inductive nor a capacitive effect. When a sinusoidal p.d. is applied across a pure resistor, the current at any instant in time will obey Ohm's law.

If i is the instantaneous value of current, then $i = e/R$. Figure 19.12 shows how the waveforms will look.

As the waveforms of the p.d. of the resistor and the current in the resistor rise and fall together, the two waveforms are said to be **in phase**.

Example 19.6 An alternating p.d. of peak value 5 V and frequency 1 kHz is applied across a resistor of 20 Ω. Draw, on the same axes, the waveforms of the p.d. and current.

Peak value of current

$$I_m = \frac{V_m}{R} = \frac{5}{20} = 0.25 \text{ A (positive and negative values)}$$

(See points A on Fig. 19.13 which shows the waveforms.)
When $e = -2$ V, $I = -2/20 = -0.1$ A (see points B on Fig. 19.13).
When $e = 0$, $I = 0/20 = 0$ (see points C on Fig. 19.13).

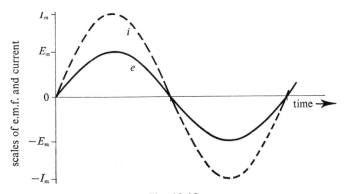

Fig. 19.12

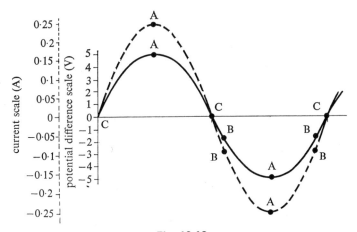

Fig. 19.13

Question 19.6 A pure resistor of 150 Ω is supplied with an alternating current of peak value 2 A and frequency 200 Hz. Draw, to scale, the waveforms of current and p.d. across the resistor. Calculate the period of the waveforms. (5 ms)

19.2 The motor effect

If instead of rotating the loop of wire by external means a current is fed into the loop, as shown by Fig. 19.14, then when the loop is in position X–X, a force will be caused on the loop which will tend to make it rotate.

The cause of the force was explained in Chapter 17, Section 17.7. As an alternative to the method shown in Chapter 17 for finding the direction of the force on a conductor, **Fleming's left-hand rule** may be used:

The thumb, first, and second fingers of the left hand are held at right angles to each other. As shown in Fig. 19.15 they will then represent, respectively, the direction of force and possibly movement, the direction of the field, and the direction of current.

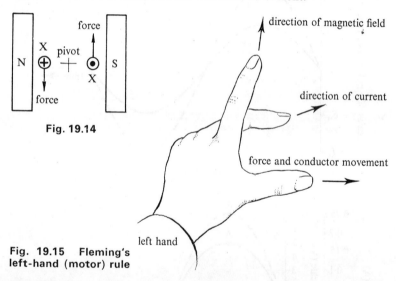

Fig. 19.14

Fig. 19.15 Fleming's left-hand (motor) rule

Until the method shown in Chapter 17, Section 17.7, is fully understood it is suggested that Fleming's left-hand rule be used as a check.

Question 19.7 Apply Fleming's left-hand rule to check the direction of forces shown in Fig. 19.14.

Torque produced by the motor effect

As the forces on each side of the loop are in opposite directions, the torque caused by the reaction of the fields will be in the form of a couple.

In Chapter 5, Section 5.7, the turning moment or torque of the loop was shown to be the product of one of the forces and the distance between the forces.

Example 19.7 If the current in the loop is 5 A and the magnetic field has a density of 0·4 T, find the torque acting on the coil. The radius of rotation is 2 cm and the effective length of each side of the loop within the field is 5 cm.

From $F = BIl$
$= 0·4 \times 5 \times 5 \times 10^{-2}$ N/conductor
$= 0·1$ N/conductor
Torque $= F \times$ Diameter $= 0·1 \times 2 \times 2 \times 10^{-2}$
$= 4$ millinewton-metres

Question 19.8 A coil of 10 turns of wire is wound on a square former of sides 3 cm. The coil is positioned so that its conductors are at right angles to a field of flux density 1·2 T, all the coil being within the field. If the coil is carrying a current of 15 A, find (a) the force acting on each conductor and (b) the torque produced by the coil. (0·54 N, 0·162 N m)

If the torque is sufficient to turn the loop the rotation will cease when the loop has rotated through 90°. The forces caused in this position will be as shown in Fig. 19.16. From this diagram the forces will be seen to be pulling the loop apart and not providing a torque.

The torque will fall from a maximum at position *X–X*, Fig. 19.14, to zero at position *Y–Y*, Fig. 19.16, the shape of the fall being that of the section of the sine wave shown in Fig. 19.17.

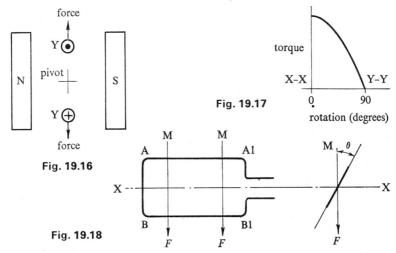

Fig. 19.16

Fig. 19.17

Fig. 19.18

Additional questions

19.9 Draw a graph of the waveform $e = 16 \sin \theta$ from $\theta = 180°$ to $\theta = 540°$. From the graph find the value of e.m.f. at (a) $\theta = 210°$, (b) $\theta = 240°$, and (c) $\theta = 450°$.

19.10 (a) Define the terms 'frequency' and 'period'.
(b) An e.m.f. has a frequency of 5 kHz. Find the period.

19.11 Explain what is meant by the peak value of an alternating e.m.f.
An e.m.f. has a peak value of 100 V and frequency 400 Hz. Sketch the waveform for two periods and find (a) the time for one period, (b) the time taken for the e.m.f. to rise from zero to 100 V, and (c) the time for the e.m.f. to fall from 100 V to 50 V.

19.12 An e.m.f. of peak value 250 V at a frequency of 50 Hz is applied across a pure resistor of 50 Ω. Draw, on the same graph, waveforms representing the e.m.f. and current. Find (a) the peak value of current and (b) the time taken for the current to reach its peak value from zero.

19.13 A coil of 25 turns is wound on a square former of sides 5 cm and is situated completely within a magnetic field of density 0·2 T. The position of the coil is as the loop in Fig. 19.1. If the current flowing in the coil is constant at 8 A, draw, to scale, a graph to show how the torque of the coil varies as the coil rotates. State the limit of its self-rotation.

19.14 A straight wire stretched between two points has a resistance of 10 Ω and carries an alternating current of peak value 2 A and frequency 50 Hz. Sketch a graph to show how the voltage across the wire varies with time during an interval of 0·04 s.
What would you expect to happen if (a) one end of a strong magnet were held close to the wire, and (b) the wire passed along the axis of a solenoid carrying a direct current? (C.G.L.I.)

19.15 An e.m.f. is generated in coil A_1 AB B_1 by rotation about axis XX in magnetic field MF (Fig. 19.18). State TWO values of angle θ that will give maximum generated e.m.f. and TWO values that will give zero e.m.f. Give reasons. (C.G.L.I.)

19.16 A rectangular coil similar to that in Fig. 19.18 has sides AA_1 = 8 cm and AB = 5 cm. There are 20 turns of wire carrying a current of 0·3 A and the magnetic field MF has flux density 0·5 T.
(a) Calculate the maximum torque tending to turn the coil about axis XX and state the value of angle θ to give this maximum. (The units of torque must be stated.)
(b) Say how the torque will vary as θ increases from 0 to 90°, and explain why. (C.G.L.I.)

19.17 Draw to scale the waveform of a sinusoidal voltage, which rises from zero to a maximum value of 200 V in a time of 0·0025 s. One complete cycle is to be drawn. Determine either from your graph or by calculation (a) the frequency, (b) the time taken to rise from zero to a value 120 V, (c) the further time interval to the next occurrence of the value of 120 V, and (d) the value of the voltage at 0·001 s after maximum negative value. (C.G.L.I.)

19.18 An alternating voltage of amplitude 10 V and frequency 100 Hz is applied across a 1000 Ω resistor. Sketch graphs to show how the current through the resistor and the voltage across it vary with time over a period of 20 ms.
Describe one method by which such a voltage may be generated.
(C.G.L.I.)

Answers to Additional Questions

(Other answers are to be found at the end of the questions.)

CHAPTER 1
1.7 44·1 m; 29·4 m/s
1.8 2 m/s; 0·4 m/s²
1.9 5 s
1.10 98·48 m; 44·2 m/s
1.11 40·2 m; 211·4 m
1.12 2·233 m/s²; 0·925 m/s²
1.13 87·3 cm³
1.14 0·83 kg
1.15 2·56; aluminium

CHAPTER 2
2.4 Accelerating (starting to descend); decelerating (stopping)
2.5 6000 N
2.7 6·88 N; 0·7 kgf
2.8 0·5886 N; 60 gf
2.9 20 kg m/s; 2 × 10⁶ g cm/s
2.10 25 N
2.11 Decelerating; 1·471 m/s²
2.12 5·14 m/s²
2.13 120 m/s; 50 N; 100 N
2.14 98·1 N; 740·5 N; 9·81 m/s²

CHAPTER 3
3.3 0·0006; tensile
3.8 31·8 MPa
3.9 63·6 MPa
3.10 5 mm
3.11 153·5 × 10⁹ Pa
3.12 37·7 kg
3.13 605 kgf or 5·93 kN
3.14 5 230 000 Pa or 53·2 kgf/cm²
3.15 1·59 mm

CHAPTER 4
4.10 20·7 N; 6·2 kgf; 16·5 N
4.11 14·5 N; 4·4 kgf; 7·4 N
4.12 8·5 kgf; 5 kgf
4.13 Tie; 11 tonnef; jib 18 tonnef
4.14 (a) 13 kgf, 22 kgf; (b) 26 kgf, 31·5 kgf, 22 kgf; (c) compression, compression, tension
4.15 (a) 34 kgf, 11 kgf; (b) 39 kgf and 22·5 kgf compression, 19·25 kgf tension
4.16 66 kgf; 25 kgf; 44 kgf; 50 kgf
4.17 57 kgf tension; 28·5 kgf compression
4.18 74 kgf; 52 kgf
4.19 1 N horizontal; 1·73 N vertical; 3·39 N; 1·695 m/s²
4.20 10 kgf; 199·75 kgf; 0·4905 m/s²; less than 200 kg

CHAPTER 5
5.9 920 N
5.10 150 N; 450 N
5.11 7·07 kgf
5.12 707 gf; 6·93 N
5.13 4 N; 6 N
5.14 6·5 N; 8·5 N
5.15 7·143 kgf; 2·857 kgf
5.16 6·66 kgf; 0; 16·66 kgf
5.17 0·3 N m; 0·4242 N m
5.18 1·29 kgf; 2·31 kgf
5.19 75·6°; 32·4° twist
5.20 Greater than 10 kgf; 120 kgf; 15 kgf

CHAPTER 6
6.12 50 kcal; 209 350 J
6.13 49 050 J
6.14 1 962 000 J; 1·09 kW; 344 000 J
6.16 6·125 kW; 0·136 kWh

6.17 2015 J; 1325 J; 3340 J
6.18 784·8 J; 450 J; 334·8 J; 57·4 per cent
6.19 32 min 40 s; 1·25 kW
6.20 100 J, 50 W; 1885 J, 31·42 W; 32 700 J, 109 W

CHAPTER 7
7.11 Worm and wheel; 40; 53·3 per cent
7.12 55·5 kgf
7.13 3; 52 320 J; 13 080 J
7.14 6; 83·3 per cent; 73·5 W
7.15 6; 4·8; 80 per cent
7.16 80 per cent
7.17 2·25; 40·5 N
7.18 30; 11·1 N m
7.19 63·8 kgf; 93·8 N m
7.20 40; 1790 J; 7·1 N m

CHAPTER 8
8.6 122·7 N m; 0·173 hp
8.7 (b) or (c)
8.8 16·67 N m; 0·141 hp
8.9 3·48 rad/s; 50·46 cm/s; 22·62 cm/s
8.10 61·1 rad/s; 83·3 rad/s; 0·74 rad/s²

CHAPTER 9
9.5 1·71 cm
9.6 131·5°C
9.7 4·19 cm³; 4·167 cm²; 1·9994 cm
9.8 0·938 m³
9.9 37·55 cm

CHAPTER 10
10.10 1·6 kgf/cm² absolute; 0·566 kgf/cm² gauge
10.11 −126·5°C; 459·5°C
10.12 0·883 kgf/cm² absolute; 86 500 Pa absolute
10.13 1·94 kgf/cm²
10.14 103·2°C
10.15 50 244 J
10.16 211 025 J; 3 349 600 J; 2 093 500 J; 5 654 125 J
10.17 0·9528p
10.18 0·5174 m³
10.19 389 kgf; 5·98 kgf/cm² absolute; 4·947 kgf/cm² gauge
10.20 342 K or 69°C

CHAPTER 11
11.4 Answer (b)
11.5 The proton (positive charge) attracts the electron (negative charge)
11.6 A compound is constructed from more than one type of atom. An element is constructed from only one type of atom
11.7 Copper and P.V.C. Copper electrical conductor. P.V.C. electrical insulator
11.8 50 V
11.9 17·5 s
11.10 2 kΩ
11.11 20 s

CHAPTER 12
12.13 1 A; 2 V, 3 V, 5 V
12.14 1·33 Ω
12.15 1·2 Ω; 1·5 A, 1 A, 2·5 A
12.16 0·5 A; 1 V, 5 V, 4 V, 2 V
12.17 (a) 0·89 A, 0·67 A, 0·44 A; (b) 4 Ω
12.18 8 Ω; 2 A, 1·33 A, 0·67 A; 6 V, 9 V
12.19 125 Ω
12.20 0·576 W, 0·192 W, 0·192 W, 2·24 W
12.21 (a) 0·6 W; (b) 0·2 A; 0·025 A, 0·075 A, 0·1 A; (c) 4 C
12.22 (a) 200 V, 4 A; (b) 0·16 kW; (c) 0·96 kW; (d) 0·591 hp
12.23 (a) 18 V; (b) 4 A; (c) 0·5 Ω; (d) 2160 J; (e) 64 W; (f) 36 A
12.24 (a) 500 W; (b) 3 kWh
12.25 (a) 1·49 A, 1·86 A, 1·86 A; (b) 22·3 V, 13 V, 9·3 V
12.26 Current at A, 0·05 A → Current at C, 0·05 A ← Current at B, 0·1 A → Current at D, 0·2 A ← 6 V battery, between A–B, disconnected

CHAPTER 13
13.11 10·43 Ω; 10·86 Ω; 11·29 Ω
13.12 (a) 2750 Ω; (b) 1 A, 0·091 A
13.13 15·18 Ω single; 30·36 Ω loop
13.14 −0·000 267 Ω/Ω at 0°C/°C
13.15 14 Ω
13.16 157 Ω
13.17 0·472 Ω; 7·552 Ω
13.18 22·8 m; 16 200 J
13.19 0·004 07 Ω/Ω at 0°C/°C
13.20 4·4 Ω

13.21 4·5 Ω
13.22 14·32 Ω
13.23 0·867 Ω; 0·802 Ω; 0·963 Ω

CHAPTER 14
14.8 25 Ω; 3·94 Ω; 0·757 Ω
14.9 4850 Ω; 19 850 Ω; 49 850 Ω; 199 850 Ω; 499 850 Ω
14.10 ∞; 900 Ω; 150 Ω; 0
14.11 0·5 MΩ; 50 per cent error
14.12 1·127 A; 0·202 A; 0·202 A; 0·104 A; 1·25 A
14.13 400 Ω; 1200 Ω; 6·09 Ω
14.14 3960 Ω; 11 960 Ω; 39 960 Ω; 119 960 Ω
14.15 0·050 05 Ω; 1450 Ω; 950 Ω; 808 Ω
14.16 True 50 V; 33⅓ V; 3⅓ mA; 10 kΩ; 50 per cent
14.17 2·22 Ω; 6·67 Ω; 20 Ω; 60 Ω; 19 980 Ω
14.18 9·1 A

CHAPTER 15
15.9 2·98 g; 0·376 g
15.10 23 h 40 min
15.11 262 A
15.12 8 h 45 min
15.13 1 h 6 min 40 s
15.15 2·4 mA Leclanché would do; 3·23 A lead–acid or alkaline required.
15.17 4 h 8 min
15.18 2·015 g; 1·12 mg/C; 10 V lead–acid
15.19 0·553 mg/C; 23 min; 0·199 g

CHAPTER 16
16.2 200 cm²
16.3 (a) and (c) no effect; (b) flux density increased
16.4 To complete the magnetic circuit with a low-reluctance path. The molecules will then be held in a magnetized position
16.5 20 mWb
16.6 3·33 T; 10 T
16.8 Store east to west to prevent hard steel scribers becoming magnetized

CHAPTER 17
17.9 5 mA
17.10 59·8 At/m
17.11 0·3142 mT
17.12 Increased by a factor of 500, that is 0·1571 T
17.13 2·514 mWb
17.14 1·87 × 10⁶ At/Wb
17.15 93·75 μWb
17.16 40 cm; 1·509 T
17.17 4·14 × 10⁻⁸ Wb
17.18 (a) attract; (b) repel
17.20 0·3125 N; 0·9375 N cm; 90°
17.21 2·18 T

CHAPTER 18
18.6 20·8 V
18.7 25 V
18.8 10 A
18.9 0·08 ms
18.12 E.m.f. reversed; trebled; halved; zero; current caused

CHAPTER 19
19.9 −8 V; −13·82 V; 16 V
19.10 0·2 ms
19.11 2·5 ms; 0·625 ms; 0·417 ms
19.12 5 A; 5 ms
19.13 0·1 N m maximum; 90°
19.14 (a) vibrate; (b) nothing
19.15 0°; 180°; 90°; 270°
19.16 1·2 N cm; 0°; 180°; by law: Torque = $T_m \sin \theta$
19.17 100 Hz; 1·023 ms; 2·954 ms −161·8 V
19.18 $I_m = 10$ mA; periodic time = 10 ms

Index

Absolute pressure, 98
Absolute temperature scales, 59, 97
Acceleration, 3
Acceleration due to gravity, 11
Accumulator, 158, *see also* Secondary cell
Addition of force vectors, 24
Alkaline secondary cells, 161
Alternating current, 193
A.C. circuit with pure resistance, 198
Alternating e.m.f., 193
Amalgamated zinc, 158
Ammeter, 116, 143
Ammeter multiplier, 144
Ammeter shunt, 143
Ampere, 112; S.I. unit, 181
Ampere-hour, 160
Angle, 86
Angle of friction, 39
Angular acceleration, 89
Angular velocity, 87
Anode, 154
Arc at switch contacts, 190
Atmospheric pressure, 98
Atom, 107
Atomic number, 108
Atomic structure, 107

Back e.m.f., 189
Battery, 156; *see also* Primary and Secondary cells
Beams, simply supported, 50
Bell-cranked lever, 71
Bimetallic strip, 94
Boiling point, 103
Bow's notation, 26
Boyle's law, 99
Breaking stress, 17

Calorie, 59
Capacitor, 111
Capacity, 111
Cathode, 154
Cell, electric, 115, 156; charging of, 125
Cells in parallel, 126; in series, 124; internal resistance of, 124

Celsius temperature scale, 59, 98
Centigrade temperature scale, 59, 98
Centre of gravity, 30
Centre of mass, 30, 50
Characteristic equation of a gas, 100
Charge, electric, 106; quantity of, 110
Charging of electric cells, 125, 158
Charles' law, 98
Chemical effect of electric current, 154
Circuit diagram, 115
Coefficient of expansion, 95; of friction, 13
Coherent system of units, 1
Components of a force, 37
Compound lever, 71
Compounds, 106
Compression, 16
Concurrency, 30
Conductance, 135
Conduction of heat, 104
Conductivity, 136
Conductors, electric, 109, 112
Conservation of energy, 59
Constant volume relationship, 100
Convection of heat, 104
Convectional current flow, 112
Corrosion, electrolytic, 164
Cost of energy, 65
Coulomb, 110; S.I. unit, 112
Couple, 55
Critical angle of gradient, 39
Cubical expansion, 93
Current, electric, 111; S.I. unit of, 181
Cycle, 196

Damping of moving coil meter, 191
Deceleration, 3
Density, 6
Depolarizer, 157
Derived units, 2
Differential axle, 77
Distance–time graph, 3
Double shear, 16
Dry cell, 157

Eddy current, 191
Effect of voltmeter on a circuit, 145

INDEX 207

Efficiency, 63, 69
Effort, 69
Elastic limit, 19
Elasticity, Young's modulus, 19
Electric cell, 115, 156
Electric charge, 106
Electric circuit, 115
Electric current, 111
Electric field, 109
Electric motor effect, 200
Electrical conductor, 109
Electrical energy, 59
Electrical insulator, 109
Electrical power, 117
Electrical resistance, 112, 116
Electrochemical equivalent, 156
Electrode, 154
Electrolysis, 154
Electrolyte, 154
Electrolytic cell, 154
Electrolytic conductor, 154
Electrolytic corrosion, 164
Electromagnet, 175
Electromagnetic induction, 183
Electromagnetism, 173
Electromotive force, 109
Electron, 108; mass of, 108
Electron current flow, 112
Electroplating, 162
Electroscope, 106
Elements, 107
Energy, 58, 101, 113; cost of, 65; forms of, 59, 101
Equilibrant force, 25
Equilibrium, 12, 25, 46
Expansion, cubical, 93; linear, 92; of gases, 97; of liquids, 93; of solids, 93; superficial, 93
Extension of springs, 20

Factor of safety, 21
Fall of potential, 119
Farad, 111
Faraday's law of electromagnetic induction, 184
Faraday's laws of electrolysis, 155
Fauré process, 160
Ferromagnetic materials, 178
Field, electric, 109
Field, magnetic, 169; of conductor, 173; of solenoid, 175
Fleming's left-hand rule, 200; right-hand rule, 196
Flux, magnetic, 169
Flux density, magnetic, 169
Flywheel, 90

Force, 9; on conductor in magnetic field, 179; on inclined plane, 38; units of, 10, 11
Forces board, 29
Forms of energy, 58, 101
Frequency, 196
Friction, coefficient of, 13; force, 13; and the inclined plane, 38, 73

Galvanometer, 183
Gas laws, 98
Gauge pressure, 98
Gear box, 80
Gearing, 79
Generation of e.m.f., 183, 193
Gradient of graph, 4
Gravitational force, 11
Gravity, acceleration due to, 6, 11

Heat energy, 59, 113
Heat sink, 104
Heat transfer, 104
Heating effect of electric current, 113
Henry, 189
Hertz, 196
Hooke's law, 19
Horsepower, 64
Hydroelectric scheme, 59

Ideal gas, 97
Inclined plane, 38, 62, 73
Induced current, 183
Induced e.m.f., 183, 193
Inductance, mutual, 188; self, 189
Inertia, 9, 90
Input, energy, 63; power, 64
Insulators, 109
Internal resistance of cell, 123
Inverse square law (magnetism), 167
Ion, 154
Ionic, 154

Jib crane, 30
Joule, 58
Joule's equivalent, 59

Kelvin scale of temperature, 59, 97
Kilocalorie, 59
Kilogramme, 1
Kilogramme-force, 11
Kilowatt, 64
Kilowatt-hour, 64
Kinetic energy, 60, 90

Latent heat, 103
Law of inertia, 9; of momentum, 9; of reaction, 12; sine wave, 195
Lead-acid cell, 159

Leclanché cell, 157
Lenz's law, 189
Lever, 69
Lever systems, 71
Leverage, 70
Limit of proportionality, 19
Linear expansion, 92
Load, 69
Local action, 158
Loop resistance, 132

Machine, 69
Magnet, 168
Magnetic circuit, 171
Magnetic effect of current, 173
Magnetic field, 168; direction, 168, 173; intensity, 177; patterns, 168, 174
Magnetic flux, 169; density, 169
Magnetic materials, 179
Magnetism, 167
Magnetomotive force, 175
Mass, 6, 23; S.I. unit of, 1
Materials, strength of, 17
Matter, properties of, 106
Maximum value of a.c., 197
Mechanical advantage, 70
Modulus of elasticity, 19
Molecular theory of magnetism, 170
Molecule, 106
Moment of a force, 44; units of, 45
Momentum, 9
Motor effect, 200
Moving coil instrument principle, 181, 191
Multiplier, voltmeter, 144
Multi-range meter, 150
Mutual inductance, 188
Mutual induction, 188

Negative resistance-temperature coefficient, 140
Neutron, 107
Newton, Sir Isaac, 9
Newton, S.I. unit of force, 10
Newton's laws of motion, 9
Nickel-cadmium alkaline cell, 161
Nickel-iron alkaline cell, 161
North pole, 167
North-seeking pole, 168
Nucleus, 107

Ohm, 116
Ohmmeter, 149
Ohm's law, 116
'Ohm's law' of magnetic circuit, 175
Ohms per volt, 148
Output power, 64
Output work, 63

Parallel circuit, 120
Parallel forces, 53
Parallelogram of forces, 25
Peak value, 197
Pendulum, 61
Period, 197
Peripheral velocity, 87
Permeability, absolute, 177; of free space, 178; relative, 178
Phase, 199
Pin-jointed framework, 32
Planté process, 160
Plastic stage, 19
Polarization, 157
Polygon of forces, 34
Positive resistance-temperature coefficient, 138
Potential difference, 110
Potential energy, 60
Potential fall, 119
Potential level, 120
Power, 63, 88, 117
Pressure, absolute, 98; atmospheric, 98; effect on boiling point, 103
Pressure cooker, 103
Pressure gauge, 98
Primary cell, 156
Properties of matter, 106
Proton, 107
Pulley systems, 74

Quantity of charge, 110; S.I. unit, 112

Radian, 86
Radiation of heat, 104
Rate of change of flux, 185
Reaction, 12, 46
Reaction force law, 12
Reference vector, 24
Relative density, 7
Relative permeability, 178
Reluctance, 171, 175
Resistance alloys, 139
Resistance, electrical, 112, 116; in a.c. circuit, 198, S.I. unit, 116
Resistivity, 131
Resistors, 118; in parallel, 120; in series, 118
Resolution of force vectors, 37
Resultant force, 24
Rotational frequency, 87
Rotating field system, 195

S.I. units, 1
Safety factor, 21
Scalar quantity, 23
Screw-jack, 74

INDEX **209**

Screw rule (electromagnetic), 173
Secondary cell, 158
Self inductance, 189
Self induction, 188
Semiconductor, 113, 140
Sensible heat, 101
Series circuit, 118
Series-parallel circuit, 126
Shear, 16
Shear force, 51
Shunt ammeter, 143
Siemen, 135
Sine wave, law, 195
Single shear, 16
Sink, heat, 104
Sinusoidal waveform, 195
Solenoid, 175
South pole, 167
South-seeking pole, 168
Space diagram, 23
Specific gravity, 7
Specific heat, 102
Specific resistance (*see* Resistivity)
Speed, 2, 23
Spring balance, 21
Spring, extension of, 20
Sliding friction, 13
Static friction, 13
Strain, 18
Strength of materials, 17
Stress, 15
Structures, effect of heat on, 96
Superconduction, 139
Superficial expansion, 93

Temperature coefficient of resistance, 137
Temperature scales, 59, 97
Tension, 16
Tesla, 169

Thermistor, 140
Thermometer, 94
Thermostat, 95
Torque, 55, 80, 89, 200
Torque converter, 80
Transfer of heat, 104
Transistor, 113
Triangle of forces, 26; applications, 30
Turning moment, 56

Ultimate strength, 19
Ultimate stress, 17

Variable force, work from, 65
Vector, 23
Vector diagram, 23
Vector quantity, 23
Velocity, 2, 23
Velocity ratio, 70
Velocity–time graph, 4
Volt, 109; S.I. unit, 117
Voltmeter, 116, 143; error caused by, 145
Voltmeter multiplier, 144

Water equivalent mass, 102
Watt, 63
Weber, 169; S.I. unit, 186
Weight, 11
Weight and inclined plane, 38
Weston reference cell, 158
Wheel and axle, 78
Winch, 81
Work, 58; by variable force, 65; diagrams, 65; on inclined plane, 62
Working stress, 21
Worm and wheel, 82

Yield point (*see* Elastic limit)
Young's modulus, 19